Developments in Electrochemistry
http://dx.doi.org/10.5772/55148
Edited by Jang H. Chun

Contributors

Yohei Okada, Kazuhiro Chiba, Yulia Budnikova, Sergei Krasnov, Virginie Lair, Hanna Ayoub, Michel Cassir, Fethi Bedioui, Sophie Griveau, Jean-Henry Calvet, Przemyslaw Sanecki, Piotr Skitał, Jang Ho Chun, Jinyoung Chun

Published by InTech

Janeza Trdine 9, 51000 Rijeka, Croatia

Notice

Publishing Process Manager Ana Pantar
Technical Editor InTech DTP team
Cover InTech Design team

Additional hard copies can be obtained from orders@intechopen.com

Developments in Electrochemistry, Edited by Jang H. Chun
p. cm.
ISBN 978-953-51-0851-1

For EU product safety concerns:
IN TECH d.o.o., Prolaz Marije Krucifikse Kozulić 3, 51000 Rijeka, Croatia,
info@intechopen.com or visit our website at intechopen.com.

Meet the editor

Professor Jang H. Chun is a Full Professor of Department of Electronic Engineering and a Dean of Graduate School in Kwangwoon University, Seoul, Republic of Korea. He holds a B.E. in Electronics from Kwangwoon University, an M.E. in Electronics from Yonsei University, Seoul, Republic of Korea, and a Ph.D. in Electrophysics from Stevens Institute of Technology, New Jersey, U.S.A. He served as a Visiting Scientist in Princeton University, U.S.A., and University of Tokyo, Japan. Professor Chun's research focuses on the development of the phase-shift method and correlation constants for determining the electrochemical Frumkin, Langmuir, and Temkin adsorption isotherms and related electrode kinetic and thermodynamic parameters at noble and highly corrosion-resistant metal and alloy/acidic and alkaline solution interfaces. He is a member of The Electrochemical Society, International Association for Hydrogen Energy, and The Korean Electrochemical Society.

Contents

Preface

An understanding of electrophysical phenomena and related electrode kinetic and thermodynamic parameters at interfaces is fundamental and necessary for developing advanced electrochemical science and technology. Observing unexpected electrophysical phenomena or developing new methods, materials, and applications at interfaces is always exciting and a great delight for electrochemists.

In both present and future, it has been clear that health, environment and energy are the most important and inescapable issues facing mankind. Developments in electrochemical methods, materials, and applications will be able to solve these issues since many developments in health treatments and environmental clean energy sources rely on the advanced electrochemical science and technology.

The book "Developments in Electrochemistry" presents current issues and developments in electrochemical methods, materials, and their applications. This book contains five feature articles in recent advanced electrochemistry. These feature articles emphasize physical phenomena rather than mathematical formalisms of electrochemistry. The topics represented in the book are: The phase-shift method and correlation constants for determining the electrochemical Frumkin, Langmuir, and Temkin adsorption isotherms at noble and highly corrosion-resistant metal and alloy/solution interfaces; Quantitative separation of an adsorption effect in the form of defined current probabilistic responses for catalyzed/inhibited electrode processes; A quick, simple, and non-invasive method to evaluate sudomotor dysfunctions; Cyclohexane-based liquid-biphasic systems for organic electrochemistry; Electrochemical transformation of white phosphorus as a way to compounds with phosphorus-hydrogen and phosphorus-carbon bonds.

I hope that this book will contribute to the understanding of the aspects of developments and issues in recent advanced electrochemistry. I also hope it can contribute to changing aspects of these developments and issues, to show that the butterfly effect really exists in electrochemistry.

I would like to thank the chapter authors, who were selected and invited by the scientific editorial board of InTech, for their feature article contributions. Also, I would like to thank the publishing process manager Ana Pantar and the technical staff of InTech for their excellent publishing assistance.

Professor Jang H. Chun, Ph.D.
Department of Electronic Engineering
Kwangwoon University
Seoul, Republic of Korea

Section 1

Developments of Electrochemical Methods and Their Applications

Chapter 1

Developments in Electrochemistry: The Phase-Shift Method and Correlation Constants for Determining the Electrochemical Adsorption Isotherms at Noble and Highly Corrosion-Resistant Metal/Solution Interfaces

Jinyoung Chun and Jang H. Chun

Additional information is available at the end of the chapter

http://dx.doi.org/10.5772/52494

1. Introduction

To obtain an environmentally clean energy source, many experimental methods have been developed and used to study the adsorption of hydrogen for the cathodic H_2 evolution reaction (HER) and hydroxide for the anodic O_2 evolution reaction (OER) at noble and highly corrosion-resistant metal and alloy/aqueous solution interfaces [1–7]. The cathodic HER is one of the most extensively studied topics in electrochemistry, electrode kinetics, renewable and sustainable energy, etc. It is well known that underpotentially deposited hydrogen (UPD H) and overpotentially deposited hydrogen (OPD H) occupy different surface adsorption sites and act as two distinguishable electroadsorbed H species, and that only OPD H can contribute to the cathodic HER [2–7]. Similarly, one can interpret that underpotentially deposited deuterium (UPD D) and overpotentially deposited deuterium (OPD D) occupy different surface adsorption sites and act as two distinguishable electroadsorbed D species, and that only OPD D can contribute to the cathodic D_2 evolution reaction (DER). However, there is not much reliable electrode kinetic data for OPD H and OPD D, i.e. the fractional surface coverage, interaction parameter, and equilibrium constant for the Frumkin adsorption isotherm, at the interfaces. Also, a quantitative relationship between the Temkin and Frumkin or Langmuir adsorption isotherms has not been developed to study the cathodic HER and DER. Thus, there is a technological need for a useful, effective, and reliable method to determine the Frumkin, Langmuir, and Temkin adsorption isotherms of OPD H and OPD D and related electrode kinetic and thermodynamic parameters. In the following discussions, H and D mean OPD H and OPD D, respectively.

Although the electrochemical Frumkin and Langmuir adsorption isotherms may be regarded as classical models and theories, it is preferable to consider the Frumkin and Langmuir adsorption isotherms for H and D rather than electrode kinetics and thermodynamics equations for H and D because these adsorption isotherms are associated more directly with the atomic mechanisms of H and D [8]. However, there is not much reliable information on the Frumkin and Langmuir adsorption isotherms of H for the cathodic HER and related electrode kinetic and thermodynamic data [1–7]. Furthermore, there is not much reliable information on the Frumkin and Langmuir adsorption isotherms of D for the cathodic DER and related electrode kinetic and thermodynamic data. Because, to the authors' knowledge, the interaction parameter and equilibrium constant for the Frumkin adsorption isotherm of H and D cannot be experimentally and readily determined using other conventional methods [3,7].

To determine the Frumkin, Langmuir, and Temkin adsorption isotherms, the phase-shift method and correlation constants have been originally developed on the basis of relevant experimental results and data. The phase-shift method is a unique electrochemical impedance spectroscopy technique for studying the linear relationship between the phase shift ($90° \geq -\varphi \geq 0°$) vs. potential ($E$) behavior for the optimum intermediate frequency (f_o) and the fractional surface coverage ($0 \leq \theta \leq 1$) vs. E behavior of the intermediates (H, D, OH, OD) for the sequential reactions (HER, DER, OER) at noble and highly corrosion-resistant metal and alloy/solution interfaces [9–29]. The θ vs. E behavior is well known as the Frumkin or Langmuir adsorption isotherm.

At first glance, it seems that there is no linear relationship between the $-\varphi$ vs. E behavior for f_o and the θ vs. E behavior at the interfaces. Thus, the tedious experimental procedures presented there [e.g. 13, 16, 19–21, 27] have been used to verify or confirm the validity and correctness of the phase-shift method. This is discussed in more detail in the section on theoretical and experimental backgrounds of the phase-shift method. However, note that many scientific phenomena have been interpreted by their behavior rather than by their nature. For example, the wave–particle duality of light and electrons, i.e. their wave and particle behaviors, is well known in science and has been applied in engineering. To explain the photoelectric effect of light, the behavior of light is interpreted as a particle, i.e. a photon, on the basis of the observed phenomena or the measured experimental data. Note that the nature of light is a wave. Similarly, to explain the tunneling effect of electrons, the behavior of electrons is interpreted as a wave on the basis of the observed phenomena or the measured experimental data. Note that the nature of the electron is a real particle, which has a negative charge and a mass. Notably, these wave and particle behaviors are complementary rather than contradictory to each other.

The comments and replies on the phase-shift method are described elsewhere [30–34]. New ideas or methods must be rigorously tested, especially when they are unique, but only with pure logic and objectivity and through scientific procedures. However, the objections to the phase-shift method do not fulfill these criteria. The objections to the phase-shift method are substantially attributed to a misunderstanding of the phase-shift method itself [27, 28]. Note especially that all of the objections to the phase-shift method can be attributed to confusion regarding the applicability of related impedance equations for intermediate frequencies and a unique feature of the faradaic resistance for the recombination step [35]. The validity and correctness of the phase-shift method should be discussed on the basis of numerical simulations with a single equation for $-\varphi$ vs. θ as functions of E and frequency (f) or relevant experimental data which are obtained using other conventional methods. The lack of the single equation for $-\varphi$ vs. θ as functions of E and f and use of incorrect values of the electrode kinetic parameters or the equivalent circuit elements for the numerical simulations given in the comments result in the confused conclusions on the phase-shift method.

In practice, the numerical calculation of equivalent circuit impedances of the noble and highly corrosion-resistant metal and alloy/solution interfaces is very difficult or impossible due to the superposition of various effects. However, it is simply determined by frequency analyzers, i.e. tools. Note that the phase-shift method and correlation constants are useful and effective tools for determining the Frumkin, Langmuir, and Temkin adsorption isotherms and related electrode kinetic and thermodynamic parameters.

This work is one of our continuous studies on the phase-shift method and correlation constants for determining the Frumkin, Langmuir, and Temkin adsorption isotherms. In this paper, as a selected example of the phase-shift method and correlation constants for determining the electrochemical adsorption isotherms, we present the Frumkin and Temkin adsorption isotherms of (H + D) for the cathodic (HER + DER) and related electrode kinetic and thermodynamic parameters of a Pt–Ir alloy/0.1 M LiOH (H_2O + D_2O) solution interface. These experimental results are compared with the relevant experimental data of the noble and highly corrosion-resistant metal and alloy/solution interfaces [11, 13, 16, 19–21, 23–29]. The interaction parameters, equilibrium constants, standard Gibbs energies of adsorptions, and rates of change of the standard Gibbs energies with θ for the Frumkin, Langmuir, and Temkin adsorption isotherms of H, D, (H + D), OH, and (OH + OD) are summarized and briefly discussed.

2. Experimental

2.1. Preparations

Taking into account the H^+ and D^+ concentrations [27] and the effects of the diffuse-double layer and pH [36], a mixture (1:1 volume ratio) of 0.1 M LiOH (H_2O) and 0.1 M LiOH (D_2O) solutions, i.e. 0.1 M LiOH (H_2O + D_2O) solution, was prepared from LiOH (Alfa Aesar, purity 99.995%) using purified water (H_2O, resistivity > 18 MΩ · cm) obtained from a Millipore system and heavy water (D_2O, Alfa Aesar, purity 99.8%). The p(H + D) of 0.1 M LiOH (H_2O + D_2O) solution was 12.91. This solution was deaerated with 99.999% purified nitrogen gas for 20 min before the experiments.

A standard three-electrode configuration was employed. A saturated calomel electrode (SCE) was used as the standard reference electrode. A platinum–iridium alloy wire (Johnson Matthey, 90:10 Pt/Ir mass ratio, 1.5 mm diameter, estimated surface area ca. 1.06 cm^2) was used as the working electrode. A platinum wire (Johnson Matthey, purity 99.95%, 1.5 mm diameter, estimated surface area ca. 1.88 cm^2) was used as the counter electrode. Both the Pt –Ir alloy working electrode and the Pt counter electrode were prepared by flame cleaning and then quenched and cooled sequentially in Millipore Milli-Q water and air.

2.2. Measurements

A cyclic voltammetry (CV) technique was used to achieve a steady state at the Pt–Ir alloy/0.1 M LiOH (H_2O + D_2O) solution interface. The CV experiments were conducted for 20 cycles at a scan rate of 200 mV · s^{-1} and a scan potential of (0 to −1.0) V vs. SCE. After the CV experiments, an electrochemical impedance spectroscopy (EIS) technique was used to study the linear relationship between the $-\varphi$ vs. E behavior of the phase shift ($90° \geq -\varphi \geq 0°$) for the optimum intermediate frequency (f_o) and the θ vs. E behavior of the fractional surface coverage ($0 \leq \theta \leq 1$). The EIS experiments were conducted at scan frequencies (f) of (10^4 to 0.1) Hz using a single sine wave, an alternating current (ac) amplitude of 5 mV, and a direct current (dc) potential of (0 to −1.20) V vs. SCE.

The CV experiments were performed using an EG&G PAR Model 273A potentiostat controlled with the PAR Model 270 software package. The EIS experiments were performed using the same apparatus in conjunction with a Schlumberger SI 1255 HF frequency response analyzer controlled with the PAR Model 398 software package. To obtain comparable and reproducible results, all of the measurements were carried out using the same preparations, procedures, and conditions at 298 K. The international sign convention is used: cathodic currents and lagged-phase shifts or angles are taken as negative. All potentials are given on the standard hydrogen electrode (SHE) scale. The Gaussian and adsorption isotherm analyses were carried out using the Excel and Origin software packages.

3. Results and discussion

3.1. Theoretical and experimental backgrounds of the phase-shift method

The equivalent circuit for the adsorption of (H + D) for the cathodic (HER + DER) at the Pt–Ir alloy/0.1 M LiOH (H_2O + D_2O) solution interface can be expressed as shown in Fig. 1a [27, 28, 37–39]. Taking into account the superposition of various effects (e.g. a relaxation time effect, a real surface area problem, surface absorption and diffusion processes, inhomogeneous and lateral interaction effects, an oxide layer formation, specific adsorption effects, etc.) that are inevitable under the experimental conditions, we define the equivalent circuit elements as follows: R_S is the real solution resistance; R_F is the real resistance due to the faradaic resistance (R_ϕ) for the discharge step and superposition of various effects; R_P is the real resistance due to the faradaic resistance (R_R) for the recombination step and superposition of various effects; C_P is the real capacitance due to the adsorption pseudocapacitance (C_ϕ) for the discharge step and superposition of various effects; and C_D is the real double-layer capacitance. Correspondingly, neither R_F nor C_P is constant; both depend on E and θ and can be measured. Note that both R_ϕ and C_ϕ also depend on E and θ but cannot be measured.

http://dx.doi.org/10.5772/52494

The numerical derivation of C_ϕ from the Frumkin and Langmuir adsorption isotherms (θ vs. E) is described elsewhere, and R_ϕ depends on C_ϕ [37,39]. A unique feature of R_ϕ and C_ϕ is that they attain maximum values at $\theta \approx 0.5$ and intermediate E, decrease symmetrically with E at other values of θ, and approach minimum values or 0 at $\theta \approx 0$ and low E and $\theta \approx 1$ and high E; this behavior is well known in interfacial electrochemistry, electrode kinetics, and EIS. The unique feature and combination of R_ϕ and C_ϕ vs. E imply that the normalized rate of change of $-\varphi$ with respect to E, i.e. $\Delta(-\varphi)/\Delta E$, corresponds to that of θ vs. E, i.e. $\Delta\theta/\Delta E$, and vice versa (see footnotes in Table 1). Both $\Delta(-\varphi)/\Delta E$ and $\Delta\theta/\Delta E$ are maximized at $\theta \approx 0.5$ and intermediate E, decrease symmetrically with E at other values of θ, and are minimized at $\theta \approx 0$ and low E and $\theta \approx 1$ and high E. Notably, this is not a mere coincidence but rather a unique feature of the Frumkin and Langmuir adsorption isotherms (θ vs. E). The linear relationship between and Gaussian profiles of $-\varphi$ vs. E or $\Delta(-\varphi)/\Delta E$ and θ vs. E or $\Delta\theta/\Delta E$ most clearly appear at f_o. The value of f_o is experimentally and graphically evaluated on the basis of $\Delta(-\varphi)/\Delta E$ and $\Delta\theta/\Delta E$ for intermediate and other frequencies (see Figs. 3 to 5). The importance of f_o is described elsewhere [21]. These aspects are the essential nature of the phase-shift method for determining the Frumkin and Langmuir adsorption isotherms.

The frequency responses of the equivalent circuit for all f that is shown in Fig. 1a are essential for understanding the unique feature and combination of (R_S, R_F) and (C_P, C_D) vs. E for f_o, i.e. the linear relationship between the $-\varphi$ vs. E behavior for f_o and the θ vs. E behavior. At intermediate frequencies, one finds regions in which the equivalent circuit for all f behaves as a series circuit of R_S, R_F, and C_P or a series and parallel circuit of R_S, C_P, and C_D, as shown in Fig. 1 b. However, note that the simplified equivalent circuits shown in Fig. 1b do not represent the change of the cathodic (HER + DER) itself but only the intermediate frequency responses.

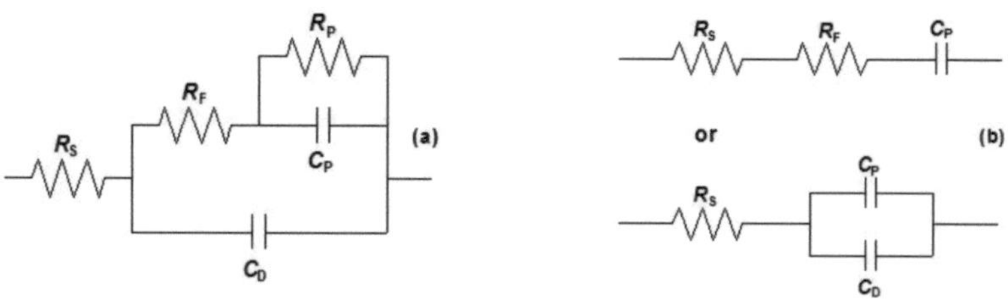

Figure 1. (a) Experimentally proposed equivalent circuit for the phase-shift method. (b) Simplified equivalent circuits for intermediate frequency responses.

At intermediate frequencies, the impedance (Z) and lagged phase-shift ($-\varphi$) are given by [27–29]

$$Z = R_S + R_F - \left(j/\omega C_P\right) \tag{1a}$$

$$-\varphi = \arctan\,[1/\omega(R_S + R_F)C_P] \tag{1b}$$

for the upper circuit in Fig. 1b or

$$Z = R_S - [j/\omega(C_P + C_D)] \tag{2a}$$

$$-\varphi = \arctan\,[1/\omega R_S(C_P + C_D)] \tag{2b}$$

for the lower circuit in Fig. 1b, where j is the imaginary unit (i.e. $j^2 = -1$) and ω is the angular frequency, defined as $\omega = 2\pi f$, where f is the frequency. Under these conditions,

$$R_P >> 1/\omega C_P \text{ and } R_P >> R_S + R_F \tag{3}$$

In our previous papers [9–24], only Eq. (1) was used with a footnote stating that C_P practically includes C_D (see Tables 1 and 2 in Ref. 20, Table 1 in Ref. 19, etc.). Both Eqs. (1) and (2) show that the effect of R_P on $-\varphi$ for intermediate frequencies is negligible. These aspects are completely overlooked, confused, and misunderstood in the comments on the phase-shift method by Horvat-Radosevic et al. [30,32,34]. Correspondingly, all of the simulations of the phase-shift method using Eq. (1) that appear in these comments (where C_P does not include C_D) [30,32,34] are basically invalid or wrong [27,28]. All of the analyses of the effect of R_P on $-\varphi$ for intermediate frequencies are also invalid or wrong (see Supporting Information of Refs. 27 and 28).

The following limitations and conditions of the equivalent circuit elements for f_o are summarized on the basis of the experimental data in our previous papers [9–29]. Neither R_S nor C_D is constant. At $\theta \approx 0$, $R_S > R_F$ and $C_D > C_P$, or vice versa, and so forth. For a wide range of θ (i.e. $0.2 < \theta < 0.8$), $R_F >> R_S$ or $R_F > R_S$ and $C_P >> C_D$ or $C_P > C_D$, and so forth. At $\theta \approx 1$, $R_S > R_F$ or $R_S < R_F$ and $C_P >> C_D$. The measured $-\varphi$ for f_o depends on E and θ. In contrast to the numerical simulations, the limitations and conditions for Eq. (1) or (2) are not considered for the phase-shift method because all of the measured values of $-\varphi$ for intermediate frequencies include (R_S, R_F) and (C_P, C_D). Correspondingly, the measured $-\varphi$ for f_o is valid and correct regardless of the applicability of Eq. (1) or (2). Both the measured values of $-\varphi$ at f_o and the calculated values of $-\varphi$ at f_o using Eq. (1) or (2) are exactly the same (see Supporting Information in Refs. 27 and 28). The unique feature and combination of (R_S, R_F) and (C_P, C_D) are equivalent to those of R_ϕ and C_ϕ. This is attributed to the reciprocal property of R_F and C_P vs. E and suggests that only the polar form of the equivalent circuit impedance, i.e. $-\varphi$ described in Eq. (1b) or (2b), is useful and effective for studying the linear relationship between the $-\varphi$ ($90° \geq -\varphi \geq 0°$) vs. E behavior at f_o and the θ ($0 \leq \theta \leq 1$) vs. E behavior. Note that the phase-shift method for determining the electrochemical (Frumkin, Langmuir, Temkin) adsorption isotherms has been proposed and verified on the basis of the phase-shift curves ($-\varphi$

http://dx.doi.org/10.5772/52494

vs. log *f*) at various *E* (see Fig. 2). The unique feature and combination of (R_S, R_F) and (C_P, C_D) vs. *E*, i.e. $-\varphi$ vs. *E* or $\Delta(-\varphi)/\Delta E$ and θ vs. *E* or $\Delta\theta/\Delta E$, most clearly appear at f_o. The linear relationship between and Gaussian profiles of $-\varphi$ vs. *E* or $\Delta(-\varphi)/\Delta E$ and θ vs. *E* or $\Delta\theta/\Delta E$ for f_o imply that only one Frumkin or Langmuir adsorption isotherm is determined on the basis of relevant experimental results (see Figs. 3 to 5). The shape and location of the $-\varphi$ vs. *E* or $\Delta(-\varphi)/\Delta E$ profile for f_o and the θ vs. *E* or $\Delta\theta/\Delta E$ profile correspond to the interaction parameter (*g*) and equilibrium constant (K_o) for the Frumkin or Langmuir adsorption isotherm, respectively. These aspects have been experimentally and consistently verified or confirmed in our previous papers [9–29].

3.2. Basic procedure and description of the phase-shift method

Note that the following description of the phase-shift method for determining the Frumkin adsorption isotherm is similar to our previous papers due to use of the same method and procedures [27,28].

Figure 2 compares the phase-shift curves ($-\varphi$ vs. log *f*) for different *E* at the Pt–Ir alloy/0.1 M LiOH (H_2O + D_2O) solution interface. As shown in Fig. 2, $-\varphi$ depends on both *f* and *E* [37–39]. Correspondingly, the normalized rate of change of $-\varphi$ vs. *E*, i.e. $\Delta(-\varphi)/\Delta E$, depends on both *f* and *E*. In electrosorption, θ depends on only *E* [40]. The normalized rate of change of θ vs. *E*, i.e. $\Delta\theta/\Delta E$, obeys a Gaussian profile. This is a unique feature of the Frumkin and Langmuir adsorption isotherms (θ vs. *E*).

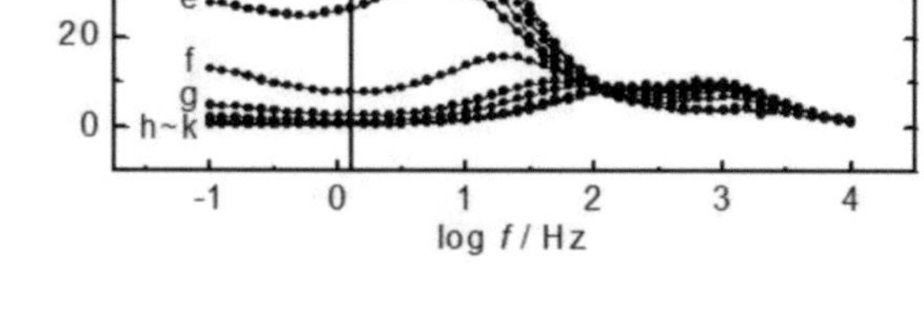

Figure 2. Comparison of the phase-shift curves ($-\varphi$ vs. log *f*) for different potentials (*E*) at the Pt–Ir alloy/0.1 M LiOH (H_2O + D_2O) solution interface. Measured values: ●. Vertical solid line: 1.259 Hz; single sine wave; scan frequency range, (10^4 to 0.1) Hz; ac amplitude, 5 mV. Dc potentials: (a) –0.659 V, (b) –0.684 V, (c) –0.709 V, (d) –0.734 V, (e) –0.759 V, (f) –0.784 V, (g) –0.809 V, (h) –0.834 V, (i) –0.859 V, (j) –0.884 V, and (k) –0.909 V (all vs. SHE).

The intermediate frequency of 1.259 Hz, shown as a vertical solid line on the $-\varphi$ vs. log *f* plot in Fig. 2, can be set as f_o for $-\varphi$ vs. *E* and θ vs. *E* profiles. The determination of f_o is experimentally and graphically evaluated on the basis of $\Delta(-\varphi)/\Delta E$ and $\Delta\theta/\Delta E$ for intermediate and other frequencies (see Figs. 3 and 4). At the maximum $-\varphi$ shown in curve a of Fig. 2, it appears that the adsorption of (H + D) and superposition of various effects are minimized; i.e. $\theta \approx 0$ and E is low. Note that θ ($0 \leq \theta \leq 1$) depends only on E. At the maximum $-\varphi$, when $\theta \approx 0$ and E is low, both $\Delta(-\varphi)/\Delta E$ and $\Delta\theta/\Delta E$ are minimized because R_ϕ and C_ϕ approach minimum values or 0. At the minimum $-\varphi$, shown in curve k of Fig. 2, it appears that the adsorption of (H + D) and superposition of various effects are maximized or almost saturated; i.e. $\theta \approx 1$ and E is high. At the minimum $-\varphi$, when $\theta \approx 1$ and E is high, both $\Delta(-\varphi)/\Delta E$ and $\Delta\theta/\Delta E$ are also minimized because R_ϕ and C_ϕ approach minimum values or 0. At the medium $-\varphi$ between curves d and e in Fig. 2, it appears that both $\Delta(-\varphi)/\Delta E$ and $\Delta\theta/\Delta E$ are maximized because R_ϕ and C_ϕ approach maximum values at $\theta \approx 0.5$ and intermediate E (see Table 1 and Fig. 4b). If one knows the three points or regions, i.e. the maximum $-\varphi$ ($\theta \approx 0$ and low E region, where $\Delta(-\varphi)/\Delta E$ and $\Delta\theta/\Delta E$ approach the minimum value or 0), the medium $-\varphi$ ($\theta \approx 0.5$ and intermediate E region, where $\Delta(-\varphi)/\Delta E$ and $\Delta\theta/\Delta E$ approach the maximum value), and the minimum $-\varphi$ ($\theta \approx 1$ and high E region, where $\Delta(-\varphi)/\Delta E$ and $\Delta\theta/\Delta E$ approach the minimum value or 0) for f_o, then one can easily determine the object, i.e. the Frumkin or Langmuir adsorption isotherm. In other words, both $\Delta(-\varphi)/\Delta E$ and $\Delta\theta/\Delta E$ for f_o are maximized at $\theta \approx 0.5$ and intermediate E, decrease symmetrically with E at other values of θ, and are minimized at $\theta \approx 0$ and low E and $\theta \approx 1$ and high E (see Table 1 and Fig. 4b). As stated above, this is a unique feature of the Frumkin and Langmuir adsorption isotherms. The linear relationship between and Gaussian profiles of $-\varphi$ vs. E or $\Delta(-\varphi)/\Delta E$ and θ vs. E or $\Delta\theta/\Delta E$ most clearly appear at f_o.

E/V vs. SHE	$-\varphi$/deg	θ [a]	$\Delta(-\varphi)/\Delta E$ [b]	$\Delta\theta/\Delta E$ [c]
−0.659	84.7	~ 0	~ 0	~ 0
−0.684	84.0	0.00830	0.08304	0.08304
−0.709	79.4	0.06287	0.54567	0.54567
−0.734	60.8	0.28351	2.20641	2.20641
−0.759	26.6	0.68921	4.05694	4.05694
−0.784	7.7	0.91340	2.24199	2.24199
−0.809	2.6	0.97390	0.60498	0.60498
−0.834	1.3	0.98932	0.15421	0.15421
−0.859	0.7	0.99644	0.07117	0.07117
−0.884	0.6	0.99763	0.01186	0.01186
−0.909	0.4	~ 1	0.02372	0.02372

Table 1. [a] ($0 \leq \theta \leq 1$) and estimated using $-\varphi$. [b] {[(neighbor phase shift difference)/(total phase shift difference)]/[(neighbor potential difference)/(total potential difference)]}. [c] {[(neighbor fractional surface coverage difference)/(total fractional surface coverage difference)]/[(neighbor potential difference)/(total potential difference)]}. Measured values of the phase shift ($-\varphi$) for the optimum intermediate frequency (f_o = 1.259 Hz), the fractional surface coverage (θ) of (H + D), and the normalized rates of change of $-\varphi$ and θ vs. E (i.e. $\Delta(-\varphi)/\Delta E$, $\Delta\theta/\Delta E$) at the Pt–Ir alloy/0.1 M LiOH ($H_2O + D_2O$) solution interface

http://dx.doi.org/10.5772/52494

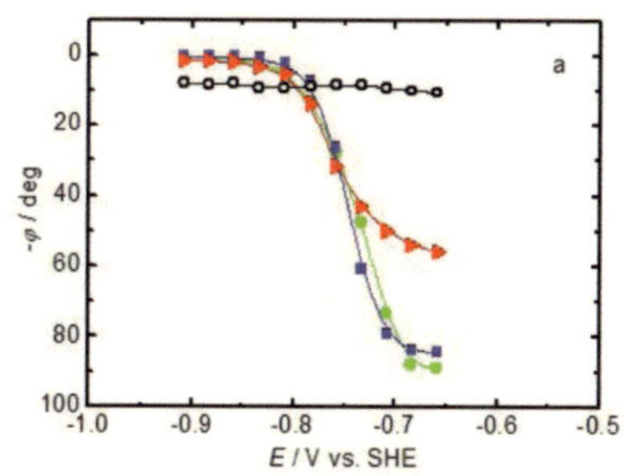

Figure 3. Comparison of (a) the phase-shift profiles ($-\varphi$ vs. E) and (b) the fractional surface coverage profiles (θ vs. E) for four different frequencies at the Pt–Ir alloy/0.1 M LiOH ($H_2O + D_2O$) solution interface. Measured or estimated values: ●, 0.1 Hz; ■, 1.259 Hz; ►, 10 Hz; ○, 100 Hz. The optimum intermediate frequency (f_o) is 1.259 Hz.

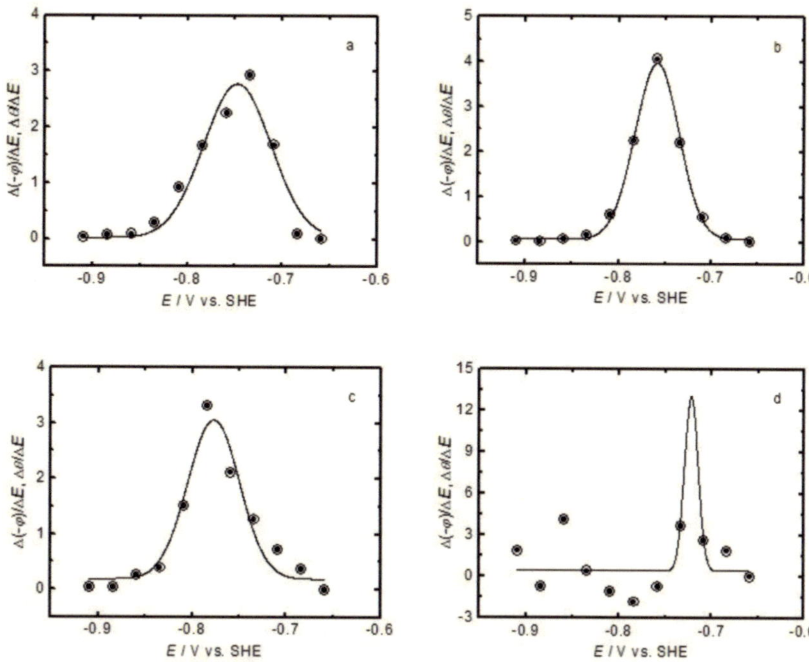

Figure 4. Comparison of the normalized rates of change of $-\varphi$ and θ vs. E, i.e. $\Delta(-\varphi)/\Delta E$ and $\Delta\theta/\Delta E$, for four different frequencies at the Pt–Ir alloy/0.1 M LiOH ($H_2O + D_2O$) solution interface. Solid curves show the fitted Gaussian profiles. Measured or estimated values: ○, $\Delta(-\varphi)/\Delta E$; ●, $\Delta\theta/\Delta E$. (a) 0.1 Hz, (b) 1.259 Hz, (c) 10 Hz, and (d) 100 Hz. The optimum intermediate frequency (f_o) is 1.259 Hz.

The procedure and description of the phase-shift method for determining the Frumkin adsorption isotherm of (H + D) at the interface are summarized in Table 1. The values of $-\varphi$ and θ as a function of E at f_o = 1.259 Hz shown in Fig. 3 are illustrated on the basis of the experimental results summarized in Table 1. The values of $-\varphi$ and θ as a function of E at f =

0.1 Hz, 10 Hz, and 100 Hz shown in Fig. 3 are also illustrated through the same procedure summarized in Table 1. However, note that the differences between the $-\varphi$ vs. E profile at f_o = 1.259 Hz and the $-\varphi$ vs. E profiles at f = 0.1 Hz, 10 Hz, and 100 Hz shown in Fig. 3a do not represent the measurement error but only the frequency response. In practice, the θ vs. E profiles at f = 0.1 Hz, 10 Hz, and 100 Hz shown in Fig. 3b should be exactly the same as the θ vs. E profile at f_o = 1.259 Hz. Because, as stated above, θ depends on only E and this unique feature most clearly appears at f_o. This is the reason why the comparison of $-\varphi$ and θ vs. E profiles for different frequencies shown in Fig. 3 is necessary to determine f_o.

The Gaussian profile shown in Fig. 4b is illustrated on the basis of $\Delta(-\varphi)/\Delta E$ and $\Delta\theta/\Delta E$ data for f_o = 1.259 Hz summarized in Table 1. Figure 4b shows that both $\Delta(-\varphi)/\Delta E$ and $\Delta\theta/\Delta E$ are maximized at $\theta \approx 0.5$ and intermediate E, decrease symmetrically with E at other values of θ, and are minimized at $\theta \approx 0$ and low E and $\theta \approx 1$ and high E. The Gaussian profiles for f = 0.1 Hz, 10 Hz, and 100 Hz shown in Fig. 4 were obtained through the same procedure summarized in Table 1. Finally, one can conclude that the θ vs. E profile at f_o = 1.259 Hz shown in Fig. 3b is applicable to the determination of the Frumkin adsorption isotherm of (H + D) at the interface. As stated above, the shape and location of the $-\varphi$ vs. E or $\Delta(-\varphi)/\Delta E$ profile and the θ vs. E or $\Delta\theta/\Delta E$ profile for f_o correspond to g and K_o for the Frumkin adsorption isotherm, respectively.

3.3. Frumkin, Langmuir, and Temkin adsorption isotherms

The derivation and interpretation of the practical forms of the electrochemical Frumkin, Langmuir, and Temkin adsorption isotherms are described elsewhere [41–43]. The Frumkin adsorption isotherm assumes that the Pt–Ir alloy surface is inhomogeneous or that the lateral interaction effect is not negligible. It is well known that the Langmuir adsorption isotherm is a special case of the Frumkin adsorption isotherm. The Langmuir adsorption isotherm can be derived from the Frumkin adsorption isotherm by setting the interaction parameter to be zero. The Frumkin adsorption isotherm of (H + D) can be expressed as follows [42]

$$[\theta/(1-\theta)]\exp(g\theta) = K_o C^+ \exp(-EF/RT) \tag{4}$$

$$g = r/RT \tag{5}$$

$$K = K_o \exp(-g\theta) \tag{6}$$

where θ ($0 \le \theta \le 1$) is the fractional surface coverage, g is the interaction parameter for the Frumkin adsorption isotherm, K_o is the equilibrium constant at g = 0, C^+ is the concentration of ions (H^+, D^+) in the bulk solution, E is the negative potential, F is Faraday's constant, R is the gas constant, T is the absolute temperature, r is the rate of change of the standard Gibbs energy of (H + D) adsorption with θ, and K is the equilibrium constant. The dimension of K is described elsewhere [44]. Note that when g = 0 in Eqs. (4) to (6), the Langmuir adsorption isotherm is obtained. For the Langmuir adsorption isotherm, when $g = 0$, the inhomogeneous and lateral interaction effects on the adsorption of (H + D) are assumed to be negligible.

At the Pt–Ir alloy/0.1 M LiOH (H_2O + D_2O) solution interface, the numerically calculated Frumkin adsorption isotherms using Eq. (4) are shown in Fig. 5. Curves a, b, and c in Fig. 5 show the three numerically calculated Frumkin adsorption isotherms of (H + D) corresponding to $g = 0$, −2.2, and −5.5, respectively, for $K_o = 5.3 \times 10^{-5}$ mol^{-1}. The curve b shows that the Frumkin adsorption isotherm, $K = 5.3 \times 10^{-5} \exp(2.2\theta)$ mol^{-1}, is applicable to the adsorption of (H + D), and Eq. (5) gives $r = -5.5$ kJ · mol^{-1}. The Frumkin adsorption isotherm implies that the lateral interaction between the adsorbed (H + D) species is not negligible. In other words, the Langmuir adsorption isotherm for $g = 0$, i.e. $K = 5.3 \times 10^{-5}$ mol^{-1}, is not applicable to the adsorption of (H + D) at the interface (see Fig. 8).

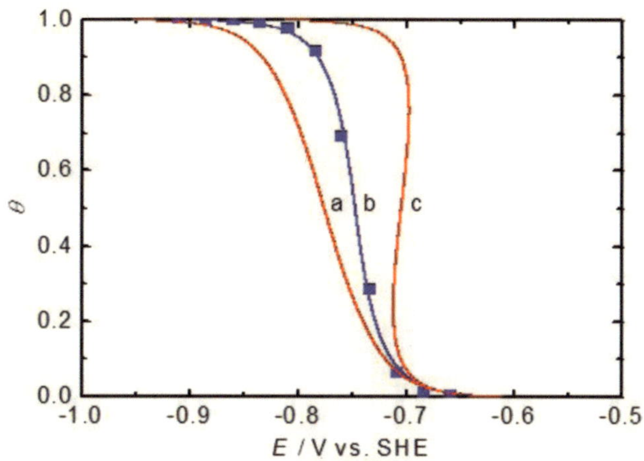

Figure 5. Comparison of the experimental and fitted data for the Frumkin adsorption isotherms of (H + D) at the Pt–Ir alloy/0.1 M LiOH (H_2O + D_2O) solution interface. Experimental data: ■. Curves show the Frumkin adsorption isotherms calculated using Eq. (4) for (a) $g = 0$, (b) $g = -2.2$, and (c) $g = -5.5$ with $K_o = 5.3 \times 10^{-5}$ mol^{-1}.

At intermediate values of θ (i.e. $0.2 < \theta < 0.8$), the pre-exponential term, $[\theta/(1 - \theta)]$, varies little with θ in comparison with the variation of the exponential term, $\exp(g\theta)$. Under these approximate conditions, the Temkin adsorption isotherm can be simply derived from the Frumkin adsorption isotherm. The Temkin adsorption isotherm of (H + D) can be expressed as follows [42]

$$\exp(g\theta) = K_o C^{+} \exp(-EF / RT) \tag{7}$$

Figure 6 shows the determination of the Temkin adsorption isotherm corresponding to the Frumkin adsorption isotherm shown in curve b of Fig. 5. The dashed line labeled b in Fig. 6

shows that the numerically calculated Temkin adsorption isotherm of (H + D) using Eq. (7) is $K = 5.3 \times 10^{-4} \exp(-2.4\theta)$ mol^{-1}, and Eq. (5) gives $r = 6.0$ kJ · mol^{-1}. The values of g and K for the Frumkin and Temkin adsorption isotherms of H, D, (H + D), OH, and (OH + OD) at the noble and highly corrosion-resistant metal and alloy/H_2O and D_2O solution interfaces are summarized in Tables 2 and 3, respectively.

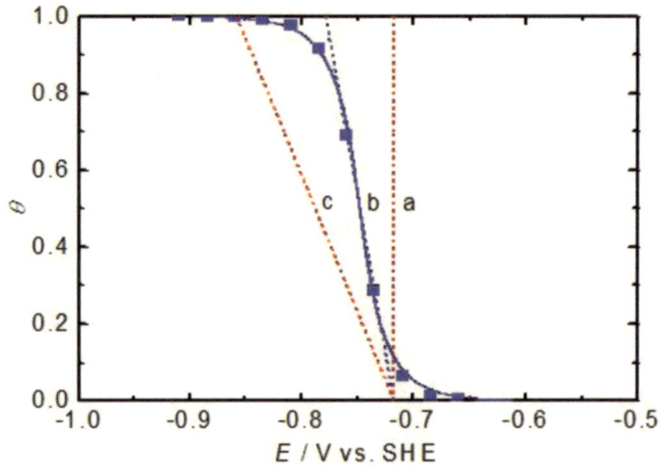

Figure 6. Comparison of the experimentally determined Frumkin adsorption isotherm and three fitted Temkin adsorption isotherms of (H + D) at the Pt–Ir alloy/0.1 M LiOH ($H_2O + D_2O$) solution interface. Experimental data: ■. The curve shows the Frumkin adsorption isotherm calculated using Eq. (4). Dashed lines show the Temkin adsorption isotherms calculated using Eq. (7) and the correlation constants for (a) $g = 0$, (b) $g = 2.4$, and (c) $g = 5.5$ with $K_o = 5.3 \times 10^{-4}$ mol^{-1}.

3.4. Applicability of the Frumkin, Langmuir, and Temkin adsorption isotherms

Figure 7 shows the applicability of ranges of θ, which are estimated using the measured phase shift ($-\varphi$) shown in Table 1, for the Frumkin adsorption isotherm at the Pt–Ir alloy/0.1 M LiOH ($H_2O + D_2O$) solution interface. Fig. 7 also shows that the phase-shift method for determining the Frumkin adsorption isotherm (θ vs. E) is valid, effective, and reasonable at $0 \leq \theta \leq 1$.

Figures 8 and 9 show the applicability of the Langmuir and Temkin adsorption isotherms at the same potential ranges, respectively. Figs. 8 and 9 also show that the Langmuir and Temkin adsorption isotherms are not applicable to the adsorption of (H + D) at the interface.

At extreme values of θ, i.e. $\theta \approx 0$ and 1, the Langmuir adsorption isotherm is often applicable to the adsorption of intermediates [42]. However, as shown in Figs. 8b and c, the validity and correctness of the Langmuir adsorption isotherm are unclear and limited even at $\theta \approx 0$ and 1. As stated in the introduction, the value of g for the Frumkin adsorption isotherm is not experimentally and consistently determined using other conventional methods. This is the reason why the Langmuir adsorption isotherm is often used even though it has the criti-

http://dx.doi.org/10.5772/52494

cal limitation and applicability. On the other hand, the Temkin adsorption isotherm is only valid and effective at $0.2 < \theta < 0.8$ (see Fig. 6). Note that the short potential range (ca. 37 mV) is difficult to observe in the Temkin adsorption isotherm correlating with the Frumkin adsorption isotherm. At other values of θ, i.e. $0 \le \theta < 0.2$ and $0.8 < \theta \le 1$, only the Frumkin adsorption isotherm is applicable to the adsorption of (H + D). Finally, one can conclude that the Frumkin adsorption isotherm is more useful, effective, and reliable than the Langmuir and Temkin adsorption isotherms at the interface.

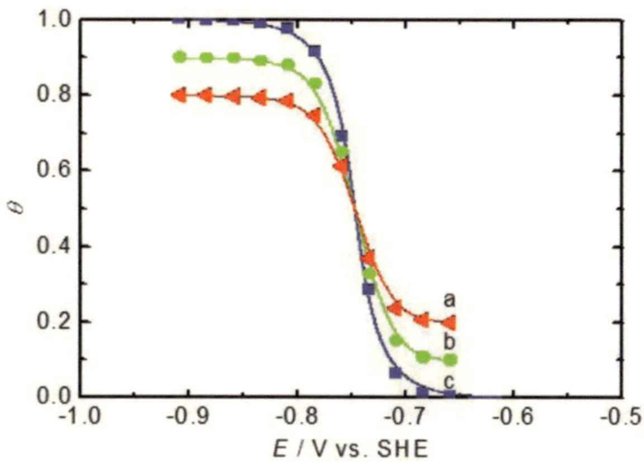

Figure 7. Comparison of ranges of θ for the Frumkin adsorption isotherm of (H + D) at the Pt–Ir alloy/0.1 M LiOH (H_2O + D_2O) solution interface. (a) $0.2 < \theta < 0.8$ (◄), (b) $0.1 < \theta < 0.9$ (●), and (c) $0 \le \theta \le 1$ (■). The blue curve is the Frumkin adsorption isotherm, $K = 5.3 \times 10^{-5} \exp(2.2\theta)$ mol^{-1}.

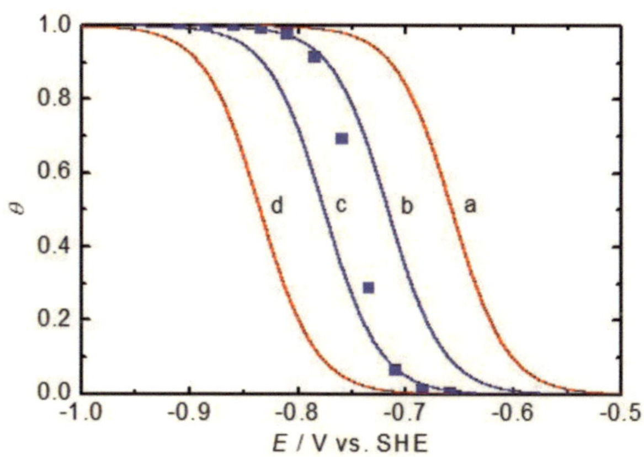

Figure 8. Comparison of the Langmuir adsorption isotherms of (H + D) at the same potential ranges. Experimental data: ■. Curves show the Langmuir adsorption isotherms (θ vs. E) calculated using Eq. (4) for $g = 0$. (a) $K = 5.3 \times 10^{-3}$ mol^{-1}, (b) $K = 5.3 \times 10^{-4}$ mol^{-1}, (c) $K = 5.3 \times 10^{-5}$ mol^{-1}, and (d) $K = 5.3 \times 10^{-6}$ mol^{-1}.

interface	adsorbate	g	K/mol^{-1}	Ref.
Pt–Ir alloy[a]/0.1 M LiOH (H_2O+D_2O)	H+D	−2.2	$5.3 \times 10^{-5} \exp(2.2\theta)$	–
Pt–Ir alloy[a]/0.1 M LiOH (H_2O)	H	−2.2	$8.6 \times 10^{-5} \exp(2.2\theta)$	27
Pt–Ir alloy[a]/0.1 M LiOH (D_2O)	D	−2.3	$2.1 \times 10^{-5} \exp(2.3\theta)$	27
Pt–Ir alloy[a]/0.5 M H_2SO_4 (H_2O)	H	−2.5	$3.3 \times 10^{-5} \exp(2.5\theta)$	28
Pt–Ir alloy[a]/0.1 M LiOH (H_2O)	OH	0.6	$5.4 \times 10^{-9} \exp(-0.6\theta)$	26
Pt–Ir alloy[a]/0.1 M LiOH (D_2O)	OH+OD	2.7	$3.9 \times 10^{-9} \exp(-2.7\theta)$	26
Pt–Ir alloy[b]/0.5 M H_2SO_4 (H_2O)	H	−2.5	$3.1 \times 10^{-5} \exp(2.5\theta)$	20
Pt–Ir alloy[b]/0.1 M KOH (H_2O)	OH	1.8	$4.7 \times 10^{-10} \exp(-1.8\theta)$	20
Pt/0.1 M KOH (H_2O)	H	−2.4	$1.2 \times 10^{-4} \exp(2.4\theta)$	25
Pt/0.5 M H_2SO_4 (H_2O)	H	−2.4	$3.5 \times 10^{-5} \exp(2.4\theta)$	21
Ir/0.1 M KOH (H_2O)	H	−2.4	$9.4 \times 10^{-5} \exp(2.4\theta)$	25
Ir/0.5 M H_2SO_4 (H_2O)	H	−2.4	$2.7 \times 10^{-5} \exp(2.4\theta)$	21
Pd/0.5 M H_2SO_4 (H_2O)	H	1.4	$3.3 \times 10^{-5} \exp(-1.4\theta)$	19
Au/0.5 M H_2SO_4 (H_2O)	H	0[e]	2.3×10^{-6}	13
Re/0.1 M KOH (H_2O)	H	0[e]	1.9×10^{-6}	16
Re/0.5 M H_2SO_4 (H_2O)	H	0[e]	4.5×10^{-7}	16
Ni[c]/0.05 M KOH (H_2O)	H	10	$1.3 \times 10^{-1} \exp(-10\theta)$	11
Ni[d]/0.1 M LiOH (H_2O)	H	7.4	$3.6 \times 10^{-4} \exp(-7.4\theta)$	29
Ni[d]/0.5 M H_2SO_4 (H_2O)	H	5.3	$4.1 \times 10^{-9} \exp(-5.3\theta)$	29
Ti/0.5 M H_2SO_4 (H_2O)	H	6.6	$8.3 \times 10^{-12} \exp(-6.6\theta)$	23
Zr/0.2 M H_2SO_4 (H_2O)	H	3.5	$1.4 \times 10^{-17} \exp(-3.5\theta)$	24

Table 2. [a] Pt–Ir (90:10 mass ratio) alloy. [b] Pt–Ir (70:30 mass ratio) alloy. [c] Ni (purity 99.994%) foil. [d] Ni (purity 99.999%) wire. [e] Langmuir adsorption isotherm. Comparison of the interaction parameters (g) and equilibrium constants (K) for the Frumkin adsorption isotherms at the noble and highly corrosion-resistant metal and alloy/H_2O and D_2O solution interfaces

interface	adsorbate	g	K/mol^{-1}	Ref.
Pt–Ir alloy[a]/0.1 M LiOH (H_2O+D_2O)	H+D	2.4	$5.3 \times 10^{-4} \exp(-2.4\theta)$	–
Pt–Ir alloy[a]/0.1 M LiOH (H_2O)	H	2.4	$8.6 \times 10^{-4} \exp(-2.4\theta)$	27
Pt–Ir alloy[a]/0.1 M LiOH (D_2O)	D	2.3	$2.1 \times 10^{-4} \exp(-2.3\theta)$	27
Pt–Ir alloy[a]/0.5 M H_2SO_4 (H_2O)	H	2.1	$3.3 \times 10^{-4} \exp(-2.1\theta)$	28
Pt–Ir alloy[a]/0.1 M LiOH (H_2O)	OH	5.2	$5.4 \times 10^{-8} \exp(-5.2\theta)$	26
Pt–Ir alloy[a]/0.1 M LiOH (D_2O)	OH+OD	7.3	$3.9 \times 10^{-8} \exp(-7.3\theta)$	26

Pt–Ir alloy[b]/0.5 M H_2SO_4 (H_2O)	H	2.1	$3.1 \times 10^{-4} \exp(-2.1\theta)$	20
Pt–Ir alloy[b]/0.1 M KOH (H_2O)	OH	6.4	$4.7 \times 10^{-9} \exp(-6.4\theta)$	20
Pt/0.1 M KOH (H_2O)	H	2.2	$1.2 \times 10^{-3} \exp(-2.2\theta)$	25
Pt/0.5 M H_2SO_4 (H_2O)	H	2.2	$3.5 \times 10^{-4} \exp(-2.2\theta)$	21
Ir/0.1 M KOH (H_2O)	H	2.2	$9.4 \times 10^{-4} \exp(-2.2\theta)$	25
Ir/0.5 M H_2SO_4 (H_2O)	H	2.2	$2.7 \times 10^{-4} \exp(-2.2\theta)$	21
Pd/0.5 M H_2SO_4 (H_2O)	H	6	$3.3 \times 10^{-4} \exp(-6\theta)$	19
Au/0.5 M H_2SO_4 (H_2O)	H	4.6	$2.3 \times 10^{-5} \exp(-4.6\theta)$	13
Re/0.1 M KOH (H_2O)	H	4.6	$1.9 \times 10^{-5} \exp(-4.6\theta)$	16
Re/0.5 M H_2SO_4 (H_2O)	H	4.6	$4.5 \times 10^{-6} \exp(-4.6\theta)$	16
Ni[c]/0.05 M KOH (H_2O)	H	14.6	$1.3 \exp(-14.6\theta)$	11
Ni[d]/0.1 M LiOH (H_2O)	H	12	$3.6 \times 10^{-3} \exp(-12\theta)$	29
Ni[d]/0.5 M H_2SO_4 (H_2O)	H	9.9	$4.1 \times 10^{-8} \exp(-9.9\theta)$	29
Ti/0.5 M H_2SO_4 (H_2O)	H	11.2	$8.3 \times 10^{-11} \exp(-11.2\theta)$	23
Zr/0.2 M H_2SO_4 (H_2O)	H	8.1	$1.4 \times 10^{-16} \exp(-8.1\theta)$	24

Table 3. [a] Pt–Ir (90:10 mass ratio) alloy. [b] Pt–Ir (70:30 mass ratio) alloy. [c] Ni (purity 99.994%) foil. [d] Ni (purity 99.999%) wire. Comparison of the interaction parameters (*g*) and equilibrium constants (*K*) for the Temkin adsorption isotherms at the noble and highly corrosion-resistant metal and alloy/H_2O and D_2O solution interfaces

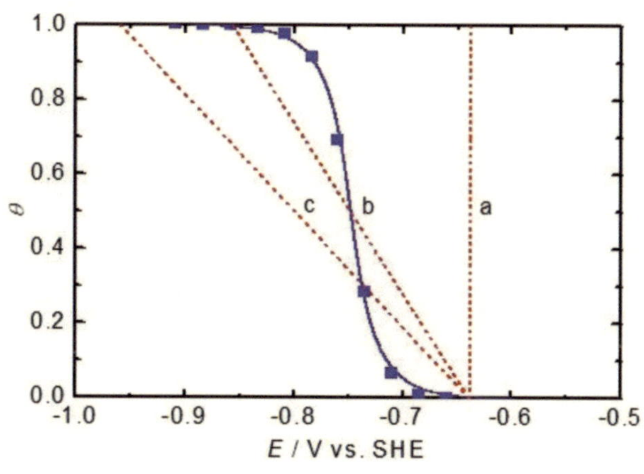

Figure 9. Comparison of the Temkin adsorption isotherms of (H + D) at the same potential ranges. Experimental data: ■. The curve shows the Frumkin adsorption isotherm calculated using Eq. (4). Dashed lines show the Temkin adsorption isotherms calculated using Eq. (7) for (a) $g = 0$, (b) $g = 8.5$, and (c) $g = 12.5$ with $K_o = 1.1 \times 10^{-2}$ mol^{-1}.

3.5. Standard Gibbs energy of adsorption

Under the Frumkin adsorption conditions, the relationship between the equilibrium constant (K) for (H + D) and the standard Gibbs energy (ΔG_θ°) of (H + D) adsorption is [42]

$$2.3RT\log K = -\Delta G_\theta^\circ \tag{8}$$

For the Pt–Ir alloy/0.1 M LiOH (H_2O + D_2O) solution interface, use of Eqs. (6) and (8) shows that ΔG_θ° is in the range ($24.4 \geq \Delta G_\theta^\circ \geq 18.9$) kJ · mol^{-1} for $K = 5.3 \times 10^{-5} \exp(2.2\theta)$ mol^{-1} and $0 \leq \theta \leq 1$. This result implies an increase in the absolute value of ΔG_θ°, i.e. $|\Delta G_\theta^\circ|$, with θ. Note that ΔG_θ° is a negative number, i.e. $\Delta G_\theta^\circ < 0$ [42]. The values of ΔG_θ° and r for the Frumkin and Temkin adsorption isotherms at the noble and highly corrosion-resistant metal and alloy/H_2O and D_2O solution interfaces are summarized in Tables 4 and 5, respectively.

4. Comparisons

4.1. Mixture solution

Curves a, b, and c in Fig. 10 show the K vs. θ behaviors of H, (H + D), and D at the Pt–Ir alloy/0.1 M LiOH (H_2O), 0.1 M LiOH (H_2O + D_2O), and 0.1 M LiOH (D_2O) solution interfaces, respectively [27]. In Fig. 10, the value of K for (H + D) is approximately equal to the average value of K for H and D isotopes. The value of K for (H + D) decreases with increasing D_2O. In other words, the value of K decreases in going from H_2O to D_2O. Over the θ range (i.e. $1 \geq \theta \geq 0$), the value of K for H is approximately 3.7 to 4.1 times greater than that for D (see Table 2). As shown in Tables 2 and 4, the values of g, K, ΔG_θ°, and r for the Frumkin adsorption isotherms of H, (H + D), and D are readily distinguishable using the phase-shift method. Fig. 10 also shows that the kinetic isotope effect, i.e. the ratio of rate constants of H and D or equilibrium constants of H and D, is readily determined using the phase-shift method (also see Table 2) [45]. Note that the kinetic isotope effect is widely used and applied in electrochemistry, surface science, biochemistry, chemical geology, physics, etc.

interface	adsorbate	ΔG_θ°/kJ·mol^{-1}	r/kJ·mol^{-1}	Ref.
Pt–Ir alloy[a]/0.1 M LiOH (H_2O+D_2O)	H+D	$24.4 \geq \Delta G_\theta^\circ \geq 18.9$	−5.5	–
Pt–Ir alloy[a]/0.1 M LiOH (H_2O)	H	$23.2 \geq \Delta G_\theta^\circ \geq 17.7$	−5.5	27
Pt–Ir alloy[a]/0.1 M LiOH (D_2O)	D	$26.7 \geq \Delta G_\theta^\circ \geq 21.0$	−5.7	27
Pt–Ir alloy[a]/0.5 M H_2SO_4 (H_2O)	H	$25.6 \geq \Delta G_\theta^\circ \geq 19.4$	−6.2	28
Pt–Ir alloy[a]/0.1 M LiOH (H_2O)	OH	$47.2 \leq \Delta G_\theta^\circ \leq 48.6$	1.5	26
Pt–Ir alloy[a]/0.1 M LiOH (D_2O)	OH+OD	$48.0 \leq \Delta G_\theta^\circ \leq 54.7$	6.7	26
Pt–Ir alloy[b]/0.5 M H_2SO_4 (H_2O)	H	$25.7 \geq \Delta G_\theta^\circ \geq 19.5$	−6.2	20

Pt–Ir alloy[b]/0.1 M KOH (H_2O)	OH	$53.2 \leq \Delta G_\theta° \leq 57.7$	4.5	20
Pt/0.1 M KOH (H_2O)	H	$22.4 \geq \Delta G_\theta° \geq 16.5$	−6.0	25
Pt/0.5 M H_2SO_4 (H_2O)	H	$25.4 \geq \Delta G_\theta° \geq 19.5$	−6.0	21
Ir/0.1 M KOH (H_2O)	H	$23.0 \geq \Delta G_\theta° \geq 17.1$	−6.0	25
Ir/0.5 M H_2SO_4 (H_2O)	H	$26.1 \geq \Delta G_\theta° \geq 20.1$	−6.0	21
Pd/0.5 M H_2SO_4 (H_2O)	H	$25.6 \leq \Delta G_\theta° \leq 29.0$	3.5	19
Au/0.5 M H_2SO_4 (H_2O)	H	32.2	0[e]	13
Re/0.1 M KOH (H_2O)	H	32.6	0[e]	16
Re/0.5 M H_2SO_4 (H_2O)	H	36.2	0[e]	16
Ni[c]/0.05 M KOH (H_2O)	H	$5.1 \leq \Delta G_\theta° \leq 29.8$	24.8	11
Ni[d]/0.1 M LiOH (H_2O)	H	$19.6 \leq \Delta G_\theta° \leq 38.0$	18.4	29
Ni[d]/0.5 M H_2SO_4 (H_2O)	H	$47.8 \leq \Delta G_\theta° \leq 61.0$	13.1	29
Ti/0.5 M H_2SO_4 (H_2O)	H	$63.2 \leq \Delta G_\theta° \leq 79.6$	16.4	23
Zr/0.2 M H_2SO_4 (H_2O)	H	$96.1 \leq \Delta G_\theta° \leq 104.8$	8.7	24

Table 4. [a] Pt–Ir (90:10 mass ratio) alloy. [b] Pt–Ir (70:30 mass ratio) alloy. [c] Ni (purity 99.994%) foil. [d] Ni (purity 99.999%) wire. [e] Langmuir adsorption isotherm. Comparison of the standard Gibbs energies ($\Delta G_\theta°$) of adsorptions and rates of change (r) of $\Delta G_\theta°$ with θ ($0 \leq \theta \leq 1$) for the Frumkin adsorption isotherms at the noble and highly corrosion-resistant metal and alloy/H_2O and D_2O solution interfaces

interface	adsorbate	$\Delta G_\theta°$/kJ·mol^{-1}	r/kJ·mol^{-1}	Ref.
Pt–Ir alloy[a]/0.1 M LiOH (H_2O+D_2O)	H+D	$19.9 < \Delta G_\theta° < 23.4$	6.0	–
Pt–Ir alloy[a]/0.1 M LiOH (H_2O)	H	$18.7 < \Delta G_\theta° < 22.2$	6.0	27
Pt–Ir alloy[a]/0.1 M LiOH (D_2O)	D	$22.2 < \Delta G_\theta° < 25.6$	5.7	27
Pt–Ir alloy[a]/0.5 M H_2SO_4 (H_2O)	H	$20.9 < \Delta G_\theta° < 24.0$	5.2	28
Pt–Ir alloy[a]/0.1 M LiOH (H_2O)	OH	$44.0 < \Delta G_\theta° < 51.8$	12.9	26
Pt–Ir alloy[a]/0.1 M LiOH (D_2O)	OH+OD	$45.9 < \Delta G_\theta° < 56.8$	18.1	26
Pt–Ir alloy[b]/0.5 M H_2SO_4 (H_2O)	H	$21.1 < \Delta G_\theta° < 24.2$	5.2	20
Pt–Ir alloy[b]/0.1 M KOH (H_2O)	OH	$50.7 < \Delta G_\theta° < 60.2$	15.9	20
Pt/0.1 M KOH (H_2O)	H	$17.8 < \Delta G_\theta° < 21.0$	5.5	25
Pt/0.5 M H_2SO_4 (H_2O)	H	$20.8 < \Delta G_\theta° < 24.1$	5.5	21
Ir/0.1 M KOH (H_2O)	H	$18.3 < \Delta G_\theta° < 21.7$	5.5	25
Ir/0.5 M H_2SO_4 (H_2O)	H	$21.5 < \Delta G_\theta° < 24.7$	5.5	21
Pd/0.5 M H_2SO_4 (H_2O)	H	$22.8 < \Delta G_\theta° < 31.8$	14.9	19
Au/0.5 M H_2SO_4 (H_2O)	H	$28.7 < \Delta G_\theta° < 35.6$	11.4	13

Re/0.1 M KOH (H_2O)	H	$29.2 < \Delta G_\theta{}^\circ < 36.0$	11.4	16
Re/0.5 M H_2SO_4 (H_2O)	H	$32.7 < \Delta G_\theta{}^\circ < 39.7$	11.4	16
Ni[c]/0.05 M KOH (H_2O)	H	$6.6 < \Delta G_\theta{}^\circ < 28.3$	36.2	11
Ni[d]/0.1 M LiOH (H_2O)	H	$19.9 < \Delta G_\theta{}^\circ < 37.8$	29.8	29
Ni[d]/0.5 M H_2SO_4 (H_2O)	H	$47.0 < \Delta G_\theta{}^\circ < 61.7$	24.6	29
Ti/0.5 M H_2SO_4 (H_2O)	H	$63.1 < \Delta G_\theta{}^\circ < 79.6$	27.8	23
Zr/0.2 M H_2SO_4 (H_2O)	H	$94.4 < \Delta G_\theta{}^\circ < 106.5$	20.1	24

Table 5. [a] Pt–Ir (90:10 mass ratio) alloy. [b] Pt–Ir (70:30 mass ratio) alloy. [c] Ni (purity 99.994%) foil. [d] Ni (purity 99.999%) wire. Comparison of the standard Gibbs energies ($\Delta G_\theta{}^\circ$) of adsorptions and rates of change (*r*) of $\Delta G_\theta{}^\circ$ with θ ($0.2 < \theta < 0.8$) for the Temkin adsorption isotherms at the noble and highly corrosion-resistant metal and alloy/H_2O and D_2O solution interfaces

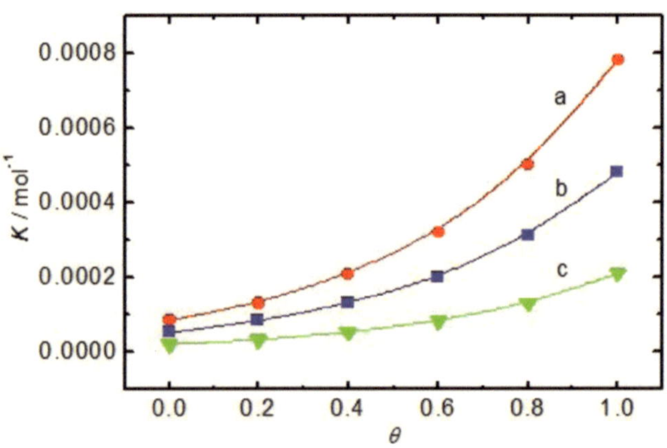

Figure 10. Comparison of the experimentally determined equilibrium constants (*K* vs. θ) for the Frumkin adsorption isotherms of H, (H + D), and D at the Pt–Ir alloy/0.1 M LiOH solution interfaces. Experimental data calculated using Eq. (6), i.e. the equilibrium constant: (a) 0.1 M LiOH (H_2O) solution (●), (b) 0.1 M LiOH ($H_2O + D_2O$) solution (■), and (c) 0.1 M LiOH (D_2O) solution (▼).

4.2. Correlation constants between the adsorption isotherms

Curves a, b, c, and d in Fig. 8 show the four numerically calculated Langmuir adsorption isotherms of (H + D) corresponding to $K = 5.3 \times 10^{-3}$, 5.3×10^{-4}, 5.3×10^{-5}, and 5.3×10^{-6} mol^{-1}, respectively. For $0.2 < \theta < 0.8$, all of the Langmuir adsorption isotherms are always parallel to each other [13,16,42]. Correspondingly, all of the slopes of the Langmuir adsorption isotherms, i.e. all of g for the Temkin adsorption isotherms, are all the same regardless of the values of K. As summarized in Tables 2 and 3, we have experimentally and consistently found and confirmed that the values of g for the Temkin adsorption isotherms are approximately 4.6 greater than those for the Langmuir adsorption isotherms, i.e. $g = 0$. Similarly, the values of g for the Temkin adsorption isotherms are approximately 4.6 greater than those for the Frumkin adsorption isotherms. Because the Frumkin adsorption isotherm is determined on the basis of the Langmuir adsorption isotherm, i.e. $g = 0$ (see Fig. 5).

In addition, we have experimentally and consistently found and confirmed that the equilibrium constants (K_o) for the Temkin adsorption isotherms are approximately 10 times greater than those (K_o or K) for the correlated Frumkin or Langmuir adsorption isotherms (see Fig. 6 and Tables 2 and 3). These factors (ca. 4.6 and 10) can be taken as correlation constants between the Temkin and Frumkin or Langmuir adsorption isotherms. The two different adsorption isotherms, i.e. the Temkin and Frumkin or Langmuir adsorption isotherms, appear to fit the same data regardless of their adsorption conditions. These aspects are described elsewhere [19, 20, 23–29].

In this work, one can also confirm that the values of g and K_o for the Temkin adsorption isotherm are approximately 4.6 and 10 times greater than those for the correlated Frumkin adsorption isotherm, respectively. The Temkin adsorption isotherm correlating with the Frumkin adsorption isotherm, and vice versa, is readily determined using the correlation constants. Note that this is a unique feature between the Temkin and Frumkin or Langmuir adsorption isotherms.

4.3. Negative and positive values of the interaction parameters for the Frumkin adsorption isotherms

A negative value of g for the Frumkin adsorption isotherm is qualitatively and quantitatively interpreted elsewhere [42,46]. Negative and positive values of g correspond to lateral attractive and repulsive interactions between the adsorbed species, respectively. At Pt, Ir, and Pt–Ir alloy/H_2O and D_2O solution interfaces, the lateral attractive interaction ($g < 0$) between the adsorbed H, D, or (H + D) species is determined [22]. As stated above, this implies an increase in $|\Delta G_\theta^\circ|$ of H, D, or (H + D) adsorption with θ ($0 \le \theta \le 1$). At Pd, Ni, Ti, and Zr/H_2O solution interfaces, the lateral repulsive interaction ($g > 0$) between the adsorbed H species is determined. This implies a decrease in $|\Delta G_\theta^\circ|$ of H adsorption with θ ($0 \le \theta \le 1$). At Au and Re/H_2O solution interfaces, the lateral interaction between the adsorbed H species is negligible, i.e. $g = 0$ or $g \approx 0$. This implies that the Langmuir adsorption isotherm is applicable. At Pt–Ir alloy/H_2O and D_2O solution interfaces, the lateral repulsive interaction ($g > 0$) between the adsorbed OH or (OH + OD) species is determined. This is significantly different from the lateral attractive interaction ($g < 0$) between the adsorbed H, D, or (H + D) species at the Pt–Ir alloy/H_2O and D_2O solution interfaces.

In contrast to Table 2, Table 3 shows that only the lateral repulsive interaction ($g > 0$) between the adsorbed H, D, (H + D), OH, or (OH + OD) species is determined. This is attributed to the values of g for the Frumkin adsorption isotherms, i.e. $g > -4.6$. Finally, one can conclude that the lateral attractive interaction ($g < 0$) between the adsorbed H, D, or (H + D) species is a unique feature of the Pt, Ir, and Pt–Ir alloy/H_2O and D_2O solution interfaces. The duality of the lateral attractive and repulsive interactions between the adsorbed H, D, or (H + D) species at the Pt, Ir, and Pt–Ir alloy interfaces is attributed to the negative values of g for the Frumkin adsorption isotherms. The Frumkin adsorption isotherm is more useful, effective, and reliable than the Temkin adsorption isotherm. As previously stated, the values of g

for the Frumkin adsorption isotherms have never been experimentally and consistently determined using other conventional methods.

4.4. Equilibrium constants

In the acidic H_2O solutions, the values of K_o, i.e. K at $g = 0$, for H at the noble metal and alloy (Pt, Ir, Pt–Ir alloy, Pd, Au, Re) interfaces are much greater than those at the highly corrosion-resistant metal (Ni, Ti, Zr) interfaces. In general, the values of K_o for H in the alkaline H_2O solutions are greater than those in the acidic H_2O solutions. The values of K_o for H in the acidic H_2O solutions are much greater than those for OH in the alkaline H_2O solutions. In the alkaline H_2O solutions, the values of K_o for H at the Ni interfaces are greater than those at the Pt, Ir, Pt–Ir alloy, Pd, Au, Re, Ti, and Zr interfaces. This is a unique feature of Ni and Ni alloy/alkaline H_2O solution interfaces. Note that Ni and Ni alloys are the metals most widely used for the cathodic HER in alkaline H_2O solutions.

The lateral interaction between the adsorbed H, D, (H + D), OH, or (OH + OD) species cannot be interpreted by the value of K_o for the Frumkin adsorption isotherm. For example, the lateral attractive interaction ($g < 0$) between the adsorbed H, D, or (H + D) species at the Pt, Ir, and Pt–Ir alloy interfaces is significantly different from the lateral repulsive interaction ($g > 0$) between the adsorbed H species at the Pd interface even though all of the Pt, Ir, Pt–Ir alloys, and Pd are the same platinum group metals and the values of K_o for H, D, and (H + D) are similar.

5. Conclusions

The Frumkin and Temkin adsorption isotherms (θ vs. E) of (H + D) and the related electrode kinetic and thermodynamic parameters (g, K, $\Delta G_\theta{}^\circ$, r) of the Pt–Ir alloy/0.1 M LiOH (H_2O + D_2O) solution interface have been determined using the phase-shift method and correlation constants and are compared with the relevant experimental data. The value of K decreases with increasing D_2O. The value of K for (H + D) is approximately equal to the average value of K for H and D isotopes. The Frumkin adsorption isotherms of H, D, and (H + D) are readily distinguishable at the interface. For $0.2 < \theta < 0.8$, the lateral attractive ($g < 0$) or repulsive ($g > 0$) interaction between the adsorbed (H + D) species appears at the interface. The Temkin adsorption isotherm correlating with the Frumkin or Langmuir adsorption isotherm, and vice versa, is readily determined using the correlation constants.

The lateral attractive interaction ($g < 0$) between the adsorbed H, D, or (H + D) species appears at the Pt, Ir, and Pt–Ir alloy interfaces. The lateral repulsive interaction ($g > 0$) between the adsorbed H species appears at the Pd, Ni, Ti, and Zr interfaces. At the Au and Re interfaces, the lateral interaction between the adsorbed H species is negligible, i.e. $g = 0$ or $g \approx 0$. The lateral repulsive interaction ($g > 0$) between the adsorbed OH or (OH + OD) species appears at the Pt–Ir alloy interfaces. The lateral attractive interaction ($g < 0$) between the adsorbed H, D, or (H + D) species is a unique feature of the Pt, Ir, and Pt–Ir alloy interfaces. For $0.2 < \theta < 0.8$, the duality of the lateral attractive and repulsive interactions between the ad-

http://dx.doi.org/10.5772/52494

sorbed H, D, or (H + D) species appears at the Pt, Ir, and Pt–Ir alloy interfaces. This unique feature of the Pt, Ir, and Pt–Ir alloy interfaces is attributed to the range of g for the Frumkin adsorption isotherms of H, D, and (H + D), i.e. $-4.6 < g < 0$.

The phase-shift method and correlation constants are the most accurate and efficient techniques to determine the Frumkin, Langmuir, and Temkin adsorption isotherms and the related electrode kinetic and thermodynamic parameters of the noble and highly corrosion-resistant metal and alloy/H_2O and D_2O solution interfaces. They are useful and effective in facilitating selection of optimal electrode materials to yield electrochemical systems of maximum hydrogen, deuterium, and oxygen evolution performances. We expect that numerical simulations with a single equation for $-\varphi$ vs. θ as functions of E and f or relevant experimental data for the phase-shift method and correlation constants will be obtained, compared, and discussed by other investigators.

Acknowledgements

The authors would like to thank Dr. Mu S. Cho (First President of Kwangwoon University, Seoul, Republic of Korea) for supporting the EG&G PAR 273A potentiostat/galvanostat, Schlumberger SI 1255 HF frequency response analyzer, and software packages. The section on theoretical and experimental backgrounds of the phase-shift method was reprinted with permission from Journal of Chemical & Engineering Data 55 (2010) 5598–5607. Copyright 2010 American Chemical Society. The authors wish to thank the American Chemical Society. This work was supported by the Research Grant of Kwangwoon University in 2012.

Author details

Jinyoung Chun[1] and Jang H. Chun[2*]

*Address all correspondence to: jhchun@kw.ac.kr

1 Department of Chemical Engineering, Pohang University of Science and Technology, Pohang, Kyungbuk, Republic of Korea

2 Department of Electronic Engineering, Kwangwoon University, Seoul, Republic of Korea

References

[1] Gileadi, E., Kirowa-Eisner, E., & Penciner, J. (1975). Interfacial electrochemistry. *Reading MA: Addison-Wesley.*

[2] Gileadi, E. (1993). Electrode kinetics. *New York: VCH.*

[3] Conway, B. E., & Jerkiewicz, G. (1995). Electrochemistry and materials science of cathodic hydrogen absorption and adsorption. *Electrochemical Society Proceedings*, 94, Pennington, NJ: The Electrochemical Society.

[4] Jerkiewicz, G., & Marcus, P. (1997). Electrochemical surface science and hydrogen adsorption and absorption. *Electrochemical Society Proceedings*, 97, Pennington, NJ: The Electrochemical Society.

[5] Jerkiewicz, G. (1998). Hydrogen sorption at/in electrodes. *Prog. Surf. Sci*, 57(2), 137-186.

[6] Jerkiewicz, G., Feliu, J. M., & Popov, B. N. (2000). Hydrogen at surface and interfaces. *Electrochemical Society Proceedings*, 2000-16, Pennington, NJ: The Electrochemical Society.

[7] Jerkiewicz, G. (2010). Electrochemical hydrogen adsorption and absorption. Part 1: Under-potential deposition of hydrogen. *Electrocatal*, 1(4), 179-199.

[8] Gileadi, E. (1967). Adsorption in electrochemistry. *Gileadi E, editor. Electrosorption. New York: Plenum Press*, 1.

[9] Chun, J. H., & Ra, K. H. (1998). The phase-shift method for the Frumkin adsorption isotherms at the Pd/H_2SO_4 and KOH solution interfaces. *J. Electrochem. Soc*, 145(11), 3794-3798.

[10] Chun, J. H., Ra, K. H., & Kim, N. Y. (2001). The Langmuir adsorption isotherms of electroadsorbed hydrogens for the cathodic hydrogen evolution reactions at the Pt(100)/H_2SO_4 and LiOH aqueous electrolyte interfaces. *Int. J. Hydrogen Energy*, 26(9), 941-948.

[11] Chun, J. H., Ra, K. H., & Kim, N. Y. (2002). Qualitative analysis of the Frumkin adsorption isotherm of the over-potentially deposited hydrogen at the poly-Ni/KOH aqueous electrolyte interface using the phase-shift method. *J. Electrochem. Soc*, 149(9), E325-330.

[12] Chun, J. H., & Jeon, S. K. (2003). Determination of the equilibrium constant and standard free energy of the over-potentially deposited hydrogen for the cathodic H_2 evolution reaction at the Pt-Rh alloy electrode interface using the phase-shift method. *Int. J. Hydrogen Energy*, 28(12), 1333-1343.

[13] Chun, J. H., Ra, K. H., & Kim, N. Y. (2003). Langmuir adsorption isotherms of over-potentially deposited hydrogen at poly-Au and Rh/H_2SO_4 aqueous electrolyte interfaces: Qualitative analysis using the phase-shift method. *J. Electrochem. Soc*, 150(4), E207-217.

[14] Chun, J. H. (2003). Methods for estimating adsorption isotherms in electrochemical systems. *U.S. Patent*, 6613218.

[15] Chun, J. H., Jeon, S. K., Kim, B. K., & Chun, J. Y. (2005). Determination of the Langmuir adsorption isotherms of under- and over-potentially deposited hydrogen for

the cathodic H_2 evolution reaction at poly-Ir/aqueous electrolyte interfaces using the phase-shift method. *Int. J. Hydrogen Energy*, 30(3), 247-259.

[16] Chun, J. H., Jeon, S. K., Ra, K. H., & Chun, J. Y. (2005). The phase-shift method for determining Langmuir adsorption isotherms of over-potentially deposited hydrogen for the cathodic H_2 evolution reaction at poly-Re/aqueous electrolyte interfaces. *Int. J. Hydrogen Energy*, 30(5), 485-499.

[17] Chun, J. H., Jeon, S. K., Kim, N. Y., & Chun, J. Y. (2005). The phase-shift method for determining Langmuir and Temkin adsorption isotherms of over-potentially deposited hydrogen for the cathodic H_2 evolution reaction at the poly-Pt/H_2SO_4 aqueous electrolyte interface. *Int. J. Hydrogen Energy*, 30(13-14), 1423-1436.

[18] Chun, J. H., & Kim, N. Y. (2006). The phase-shift method for determining adsorption isotherms of hydrogen in electrochemical systems. *Int. J. Hydrogen Energy*, 31(2), 277-283.

[19] Chun, J. H., Jeon, S. K., & Chun, J. Y. (2007). The phase-shift method and correlation constants for determining adsorption isotherms of hydrogen at a palladium electrode interface. *Int. J. Hydrogen Energy*, 32(12), 1982-1990.

[20] Chun, J. H., Kim, N. Y., & Chun, J. Y. (2008). Determination of adsorption isotherms of hydrogen and hydroxide at Pt–Ir alloy electrode interfaces using the phase-shift method and correlation constants. *Int. J. Hydrogen Energy*, 33(2), 762-774.

[21] Chun, J. H., & Chun, J. Y. (2008). Correction and supplement to the determination of the optimum intermediate frequency for the phase-shift method [Chun et al., Int. J. Hydrogen Energy 30 (2005) 247–259, 1423–1436]. *Int. J. Hydrogen Energy*, 33(19), 4962-4965.

[22] Chun, J. Y., & Chun, J. H. (2009). A negative value of the interaction parameter for over-potentially deposited hydrogen at Pt, Ir, and Pt–Ir alloy electrode interfaces. *Electrochem. Commun*, 11(4), 744-747.

[23] Chun, J. Y., & Chun, J. H. (2009). Determination of adsorption isotherms of hydrogen on titanium in sulfuric acid solution using the phase-shift method and correlation constants. *J. Chem. Eng. Data*, 54(4), 1236-1243.

[24] Chun, J. H., & Chun, J. Y. (2009). Determination of adsorption isotherms of hydrogen on zirconium in sulfuric acid solution using the phase-shift method and correlation constants. *J. Korean Electrochem. Soc*, 12(1), 26-33.

[25] Chun, J., Lee, J., & Chun, J. H. (2010). Determination of adsorption isotherms of over-potentially deposited hydrogen on platinum and iridium in KOH aqueous solution using the phase-shift method and correlation constants. *J. Chem. Eng. Data*, 55(7), 2363-2372.

[26] Chun, J., Kim, N. Y., & Chun, J. H. (2010). Determination of adsorption isotherms of hydroxide and deuteroxide on Pt–Ir alloy in LiOH solutions using the phase-shift method and correlation constants. *J. Chem. Eng. Data*, 55(9), 3825-3833.

[27] Chun, J., Kim, N. Y., & Chun, J. H. (2010). Determination of the adsorption isotherms of hydrogen and deuterium isotopes on a Pt–Ir alloy in LiOH solutions using the phase-shift method and correlation constants. *J. Chem. Eng. Data*, 55(12), 5598-5607.

[28] Chun, J., Kim, N. Y., & Chun, J. H. (2011). Determination of the adsorption isotherms of overpotentially deposited hydrogen on a Pt–Ir alloy in H_2SO_4aqueous solution using the phase-shift method and correlation constants. *J. Chem. Eng. Data*, 56(2), 251-258.

[29] Chun, J. H. (2012). Determination of the Frumkin and Temkin adsorption isotherms of hydrogen at nickel/acidic and alkaline aqueous solution interfaces using the phase-shift method and correlation constants. *J. Korean Electrochem. Soc*, 15(1), 54-66.

[30] Kvastek, K., & Horvat-Radosevic, V. (2004). Comment on: Langmuir adsorption isotherms of over-potentially deposited hydrogen at poly-Au and Rh/H_2SO_4 aqueous electrolyte interfaces; Qualitative analysis using the phase-shift method. *J. Electrochem. Soc*, 151(9), L9-10.

[31] Chun, J. H., Ra, K. H., & Kim, N. Y. (2004). Response to comment on: Langmuir adsorption isotherms of over-potentially deposited hydrogen at poly-Au and Rh/H_2SO_4 aqueous electrolyte interfaces; Qualitative analysis using the phase-shift method. *J. Electrochem. Soc* 150 (2003) E207–217. *J. Electrochem. Soc*, 151(9), L11-13.

[32] Lasia, A. (2005). Comments on: The phase-shift method for determining Langmuir adsorption isotherms of over-potentially deposited hydrogen for the cathodic H_2 evolution reaction at poly-Re/aqueous electrolyte interfaces. *Int. J. Hydrogen Energy* 30 (2005) 485–499. *Int. J. Hydrogen Energy*,30(8), 913-917.

[33] Chun, J. H., Jeon, S. K., Kim, N. Y., & Chun, J. Y. (2005). Response to comments on: The phase-shift method for determining Langmuir adsorption isotherms of over-potentially deposited hydrogen for the cathodic H_2 evolution reaction at poly-Re/aqueous electrolyte interfaces. *Int. J. Hydrogen Energy* 30 (2005) 485–499. *Int. J. Hydrogen Energy* , 30(8), 919-928.

[34] Horvat-Radosevic, V., & Kvastek, K. (2009). Pitfalls of the phase-shift method for determining adsorption isotherms. *Electrochem. Commun*, 11(7), 1460-1463.

[35] In our e-mail communications, Horvat-Radosevic et al. admitted that all of their objections to the phase-shift method in Ref. 34 were confused and misunderstood. The exact same confusion and misunderstanding about the phase-shift method also appear in Refs. 30 and 32.

[36] Gileadi, E., Kirowa-Eisner, E., & Penciner, J. (1975). Interfacial electrochemistry. *Reading, MA: Addison-Wesley*, 6.

[37] Gileadi, E., Kirowa-Eisner, E., & Penciner, J. (1975). Interfacial electrochemistry. *Reading, MA: Addison-Wesley*, 86.

[38] Harrington, D. A., & Conway, B. E. (1987). AC impedance of faradaic reactions involving electrosorbed intermediates–I. Kinetic theory. *Electrochim. Acta,* 32(12), 1703-1712.

[39] Gileadi, E. (1993). Electrode kinetics. *New York: VCH,* 291.

[40] Gileadi, E. (1993). Electrode kinetics. *New York: VCH,* 307.

[41] Gileadi, E., Kirowa-Eisner, E., & Penciner, J. (1975). Interfacial electrochemistry. *Reading, MA: Addison-Wesley,* 82.

[42] Gileadi, E. (1993). Electrode kinetics. *New York: VCH,* 261.

[43] Bockris, J., O'M Reddy, A. K. N., & Gamboa-Aldeco, M. (2000). Modern electrochemistry. *2nd edition. New York: Kluwer Academic/Plenum Press,* 2A, 1193.

[44] Oxtoby, D. W., Gillis, H. P., & Nachtrie, N. H. (2002). Principles of modern chemistry. *5th edition. New York: Thomson Learning Inc,* 446.

[45] Bockris, J., & O'M Khan, S. U. M. (1993). Surface electrochemistry. *New York: Plenum Press,* 596.

[46] Gileadi, E. (1993). Electrode kinetics. *New York: VCH,* 303.

Chapter 2

Quantitative Separation of an Adsorption Effect in Form of Defined Current Probabilistic Responses for Catalyzed / Inhibited Electrode Processes

Piotr M. Skitał and Przemysław T. Sanecki

Additional information is available at the end of the chapter

http://dx.doi.org/10.5772/53664

1. Introduction

The investigation of organic adsorbate influence on electrode process was started in works of Loshkarev e.g. [1], Schmid and Reilley [2], with theoretical formulation of Weber and Koutecký [3,4]. The problem is described in the book by Heyrovský and Kůta ([5] and literature therein). Later investigations have led to formulation of the *cup pair effect* [6], systematically developed in a series of papers e.g. [7,8]. In our earlier paper [9] it has been shown that catalytic effect of electroinactive organic adsorbate on electrode process can be isolated from CV or NPP faradaic current responses in the form of Gaussian-shaped probabilistic curves (CPR[1]). It turned out that in presence of organic adsorbate showing electrocatalytic effect, the kinetic response can be split into two defined and well shaped responses: one, due to regular reduction of cation and the second one due to reduction catalyzed by an adsorption phenomenon. The latter was named, due to its shape, the *current probabilistic response*. It has been demonstrated that CPR effect can exemplify the visualization and the quantitative measure of organic substance catalytic effect on the electrode process and simultaneously become the picture of adsorption.

The electroinhibition phenomenon was analyzed by means of selected experiments with the use of typical electroinhibitors. As an example a quasi-reversible process of Zn^{2+} electroreduction in the presence of two adsorbates such as n-alkyl alcohols [10] and two cyclodextrins was chosen [11-13,14].

1 Since PCR means Polymerase Chain Reaction and is well established in literature, in the present paper we use the abbreviation CPR.

So far, the quantitative and qualitative comparison of two systems: without adsorbate (1) and with adsorbate (2) was realized by means of four manners: dividing the adsorption non-affected and affected faradaic currents (1)/(2) or (2)/(1) [5], Tafel plots [10,12], comparison of apparent or individual rate constants [7,8] and/or simple comparison of respective curves, recorded in the presence and absence of organic substance, performed on the same plot [7].

The task of resolution of enhanced or decreased faradaic current response to show the adsorption catalytic/inhibition effect alone has not been undertaken in literature.

The literature analysis indicates that both theoretical and experimental aspects of adsorption at metal solution interface are still developed for both macroscopic and microscopic inhomogeneites (e.g. [15,16] and literature therein). Owing to existing extensive data concerning influence of organic electroinactive substances on electrode process the visualization of adsorption process in an isolated *"pure"* form is needed. It can facilitate the data processing and be competitive or parallel to capacitance currents method. Therefore, in the present paper a number of experimental facts concerning CPR is analyzed. The questions which are answered include: if the CPR approach can be applied to electrode process inhibited by organic substances and if the CPR effect is observed at solid electrodes and/or in nonaqueous media. Moreover, successful attempt to generate CPR effect by numerical simulation with the use of EE model was done.

2. Experimental

The CV, NPP and NPV experiments were carried out with the use of CV50W electrochemical analyzer (BAS, Inc.) and PGSTAT100 voltammetric analyzer (*Autolab Eco-Chemie*) for molecular oxygen reduction. The electrode system was CGME/SMDE type MF-9058 (BAS, Inc.). The mercury drop surface of working electrode was 0.0151 ± 0.0004 cm^2 for CV experiment. The glassy carbon electrodes of working surface 0.0755 ± 0.001 cm^2 (GCE3) as well as 0.0314 ± 0.001 cm^2 (GCE2) were used. For NPP experiments, the drop time was 1.0 s with a pulse-width of 50 ms and sample width 10 ms. In contrary to previous paper data, since eq. (2) contains the element of subtracting of background current, the separate background files are not necessary. The reliability of such simplification was checked by the comparison of CPR curves obtained from residual current corrected and non-corrected data. Obtained differences were not significant for the μA current scale of CPR effect.

Coulometric experiments were carried out with the use of diaphragm electrolyzer on large Hg electrode (of about 12 cm^2 surface), the volume of catholyte was 100 or 80 cm^3. A Pt counter electrode was applied in separated anodic part.

The remaining experimental details concerning DMF solutions were identical with those described in previous paper [17].

3. Kinetics

In modeling of Zn^{2+} electroreduction process the following EE sequence with two one electron steps has been taken into account:

$$Zn^{2+} \underset{k_{-1}, \beta_1}{\overset{k_1, \alpha_1}{\rightleftarrows}} Zn^{+} \underset{k_{-2}, \beta_2}{\overset{k_2, \alpha_2}{\rightleftarrows}} Zn \tag{1}$$

where: k_1, k_2, k_{-1}, k_{-2} are the heterogeneous rate constants (cm s^{-1}), α_1, α_2, and β_1, β_2 are elementary transfer coefficients of cathodic and anodic process, respectively. In the present paper the extended EE ‖ Hg(Zn) mathematical model described in [18] was applied. For the considered sequence of elementary steps (1), the reversible electron transfer for both electrochemical steps was assumed. In the extended model the considered system consist of two parts: mercury drop area from $r = 0$ to $r = R_0$ and the solution area $r > R_0$. Additionally, it has been assumed that species Zn^{2+} and Zn^{+} dissolve only in electrolyte, whereas metallic zinc dissolve and diffuse only in the mercury phase. It is caused by the fact, that deposited metallic zinc is solution-phobic and has to be immediately pulled into Hg phase to form intermetallic compound (amalgam) spontaneously [18].

The problem formulated above was solved using the *ESTYM_PDE* software. The program was designed to solve and estimate parameters of partial differential equations (PDE) describing one-dimensional mass and heat transfer coupled with a chemical reaction. The validation and comparison with other software are described in [19]. The examples of solving electrochemical problems by means of *ESTYM_PDE* software are described in previous papers [17-27]. The numerical basis of electrochemical simulation is described in references [28-33]. The two types of probabilistic fits with the use of *Origin 7.5* program were applied: the simple Gaussian model $y = y_0 + (A/(w\times(\pi/2)^{1/2}))\times\exp(-2\times((x-x_c)/w)^2)$; the asymmetric double sigmoidal model $y = y_0 + A_m\times(1/(1+\exp(-(x-x_c+w_1/2)/w_2)))\times(1-1/(1+\exp(-(x-x_c-w_1/2)/w_3)))$, where y_0, x_c, A, A_m and (w, w_1, w_2, w_3) are baseline offset, center of the peak, area, amplitude and widths of peaks, respectively.

4. Results and discussion

4.1. The systems with electrode inactive organic adsorbate

The idea [9] of such CPR plots is based on eq. (2):

$$i_{\mathrm{obs}}(E) = i_0(E) + i_{\mathrm{ads}}(E) \tag{2}$$

where, $i_0(E)$ – faradaic current due to regular (i.e. in absence of adsorbate) electrochemical response, $i_{\mathrm{ads}}(E)$ – additional faradaic current due to adsorption effect. It is assumed that the adsorption process does not change the mechanism of the electrode process, which is in accordance with the literature quoted [1-8,10-13]. In other words, the both currents in eq. (2) represent the same reduction process and the same number of electrons exchanged, moreover their final product is the same as well. The condition should be complete for each considered case.

A representative example of the resolution of deformed by inactive organic adsorbate faradaic current response into components according to eq. (2) is presented in Fig. 1. The visualization of adsorption process in its isolated *"pure"* form is regular and well-designed (Fig.1). The observed adsorption effect is catalytic since the responses (2) and (3) are enhanced and passed in relation to (1) towards positive potential.

The subtraction of experimental currents ($i_{obs}(E) - i_0(E) = i_{ads}(E)$) according to eq. (2) is reasonable since both denote the same electrode process. The applied two-electron Zn^{2+} reduction is not naturally resolved into steps since $k_2 > k_1$ [34] (cf. also the Fig. 10 caption). Therefore, an appearance of adsorption deformed responses (Fig. 1) cannot be explained in terms of resolution of coupled two-electron response. The only assumption is the additivity of currents, according to the rule commonly accepted in electrochemistry.

The two representative sets of the experimental data including five aliphatic alcohols and two cyclodextrins have been selected to establish whether the effect of inhibition can be described by current probabilistic response (CPR). In selection of inhibitors the compounds considered as simple adsorbates (a series of alcohols) as well as complex adsorbates (α,β-cyclodextrins) have been chosen. It has been shown [13] that α-cyclodextrin as inhibitor can form a condensed film. Both of the selected groups of compounds are known as electrode process inhibitors [7,10-13,35,36]. Moreover, the influence of eight additional organic adsorbates has also been investigated in connection with another electrode processes (see further).

The mechanism of Zn^{2+} electroreduction and the physical model of the inhibitors influence were proposed in quoted literature. It was shown that zinc electroreduction can be considered as successive two electron transfer steps (e.g., [37-40]). The results obtained by Fawcett and Lasia [34,41] in non-aqueous media in the presence of tetraalkylammonium perchlorates, especially in dimethylsulfooxide, revealed the complex, nonlinear potential dependence of experimental rate constant (c.f. also Fig. 1 in [41] for dimethylformamide). It indicates that the reaction consists of at least two steps. In dimethylsulfooxide the standard rate constant of the second step is greater than that of first one [34]. It means that E_0 potential of Zn^{2+}/Zn^{+} couple should be more negative than that of hypothetical Zn^{+}/Zn^{0} one. The similar conclusion is given in paper by Manzini and Lasia [40]. Additionally, it was stated that *"coulometric measurments indicate 100% yield and no metal was found in the solution after electrolysis"* [34]. It is not surprising since metallic zinc is solution-phobic and must be immediately pulled into Hg phase to form an intermetallic compound (amalgam).

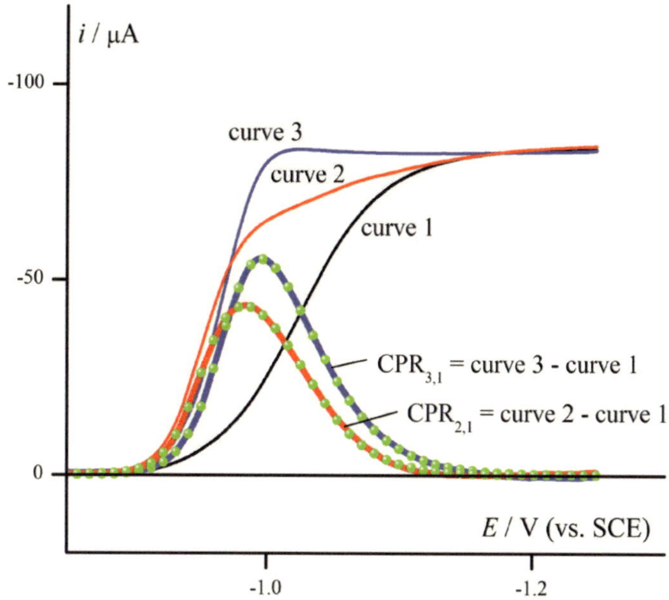

Figure 1. Experimental results expressing the idea of CPR effect. Electroreduction of Zn^{2+} in absence and presence of inert adsorbate. Regular-shaped NPV response (curve 1) and adsorption deformed NPV responses (curves 2 and 3). The difference between regular (1) and deformed response (2) or (3) is a quantitative measure of adsorption influence i.e. CPR effect: curve 3 minus curve 1 as well as curve 2 minus curve 1. Experimental conditions: NPV electroreduction of 1 mM Zn^{2+} in 1 M $NaClO_4/H_2O$ solution on Hg electrode in presence of an adsorbate: (2) – 5 mM *N,N'*-dimethylthiourea; (3) – 0.2 mM 3,4-diaminotoluene. CPR responses were approximated with the probabilistic theoretical model: points (green) are a asymmetric double sigmoidal fits. The values of parameters for $CPR_{2,1}$: $y_0 = -1.35\times10^{-7}$, $x_c = -0.989$, $A = 5\times10^{-5}$, $w_1 = 0.076$, $w_2 = 0.025$, $w_3 = 0.013$, $\chi^2 = 7.7\times10^{-14}$, $R^2 = 0.9996$ and for $CPR_{3,1}$: $y_0 = -1.54\times10^{-8}$, $x_c = -0.995$, $A = 1\times10^{-4}$, $w_1 = 0.049$, $w_2 = 0.031$, $w_3 = 0.014$, $\chi^2 = 1.3\times10^{-13}$, $R^2 = 0.9995$.

Since the pure reversible EE process of Zn^{2+} reduction is not observed in practice, there must be a reason for quasi-reversibility in form of slow step. Fawcett and Lasia [41] suggested the model in which the two step reduction is complicated by slow adsorption process of substrate and its transfer to the electrode, similarly as it was proposed for reduction of Na^+, Li^+ [42] and Ca^{2+}, Sr^{2+}, Ba^{2+} cations [43]. In the case when the dehydration desolvatation of zinc cations were considered as chemical steps [37,40], another explanation of the quasi-reversibility is provided by the presence of such preceding slow elementary processes or the situation in which electron transfer is coupled with slow desolvation.

Our results obtained in 1.0 M $NaClO_4$ solution revealed $i_p/v^{1/2} = f(v)$ dependence decreasing with scan rate v which suggests that the reduction process must be complicated by a slow step.

According to our assumption, the EE mechanism (1), i.e. the sequence of two consecutive steps $Zn^{2+} \xrightarrow{e} Zn^{+} \xrightarrow{e} Zn$, remains the same on both covered and uncovered electrode where intermediate Zn^+ is stabilized by hydration, and undergoes fast electrochemical reduction. The analogical assumptions have been undertaken in relation to Cd^{2+} electroreduction [44]. Since the CPR effect is based on comparison of the two reductions, namely in absence and presence of organic adsorbate, an extensive mechanistic analysis at the moment is not decisive for our purpose.

Our coulometric measurements were performed at sufficiently negative potentials diffusion region in order to avoid the participation of the anodic current, and gave the number of electrons exchanged 2.02 ± 0.05 for 1 mM Zn^{2+} and 1 M $NaClO_4$ solution. In the presence of two representative adsorbates the obtained results were 2.00 ± 0.09 for *n*–pentanol (0.1 M) and 1.92 ± 0.05 for *α*-cyclodextrin (0.01 M).

The representative results of applying the CPR approach to inhibition are presented in Fig. 2A and 2B. It is striking that simple, according to eq. (2), subtraction of respective parent matrices recorded in presence and absence of inhibitor (or catalyst – Fig.1), provides the curves which are elegant and defined in shape. Very similar picture is observed once convoluted CV responses are considered. The Gaussian fit done for those curves is of good or even very good quality, which justifies the use of the adjective probabilistic. None of the experimental parent curves in this paper has been idealized by smoothing or other similar procedures.

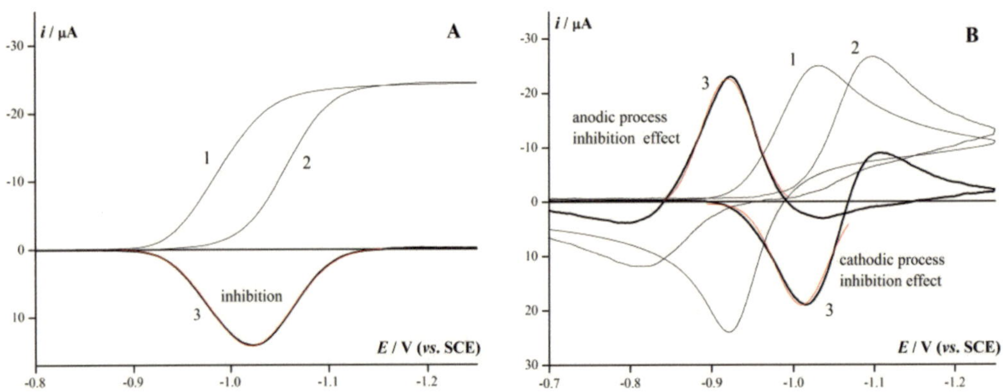

Figure 2. The separation and visualization of experimental *n*-pentanol adsorption / inhibition effect according to eq. (2). The electrode process: electroreduction of 1 mM Zn^{2+}, 1M $NaClO_4$ on Hg electrode in absence (curve 1) and in presence (curve 2) of *n*-pentanol. Curve 3 (*current probabilistic response* (CPR)) displays the inhibition effect alone in form of $i_{c,\ ads}$ (*E*). All currents are in the same scale. Red curve is a Gaussian fit. (A) NPP technique. The values of Gaussian fit parameters are: $y_0 = -1.05\times10^{-7} \pm 2.8\times10^{-8}$, $x_c = -1.0209 \pm 0.0001$, $w = 0.0844 \pm 0.0003$, $A = 1.51\times10^{-6} \pm 6.9\times10^{-9}$, $\chi^2 = 1.02\times10^{-14}$, $R^2 = 0.9996$. (B) CV technique. Scan rate 1.003 V s^{-1}. The Gaussian fit (red curves) results in good correlation parameters. The values of R^2 parameter of Gaussian fit are 0.9763 and 0.9928 for the cathodic and anodic part, respectively.

Data presented in Fig. 3 indicate the increasing influence of aliphatic alcohols on the electrode process depending on the increasing, from *n*-propanol to *n*-hexanol, length of carbon chain, similarly as it was described in the literature [10].

The similar picture is observed for two cyclodextrins (Fig. 4A,B), another typical object in investigations of electrode process inhibition [11-13].

http://dx.doi.org/10.5772/53664

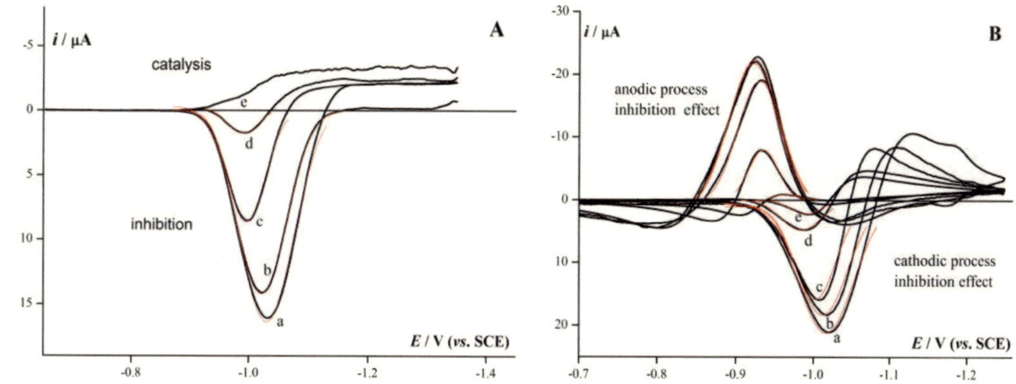

Figure 3. A comparison of experimental CPR electroinhibition effects, determined for five aliphatic alcohols from eq. (2) for electroreduction process of 1 mM Zn^{2+} in 1 M $NaClO_4$ on Hg electrode: (a) saturated ≈ 0.06 M *n*-hexanol, (b) 0.1 M *n*-pentanol, (c) 0.1 M *t*-pentanol, (d) 0.1 M *n*-butanol, (e) 0.1 M *n*-propanol. The parent NPP responses are not shown. (A) NPP technique. The Gaussian fit (red curves) results in good correlation parameters, for example the values of R^2 parameter are: (a) $R^2 = 0.9949$; (b) $R^2 = 0.9996$; (c) $R^2 = 0.9942$; (d) $R^2 = 0.9823$. (B) CV technique. Scan rate 1.003 V s^{-1}. The Gaussian fit (red curves) results in good correlation parameters for example, the values of R^2 parameter of Gaussian fit for cathodic process are: (a) $R^2 = 0.9848$; (b) $R^2 = 0.9756$; (c) $R^2 = 0.9649$; (d) $R^2 = 0.9740$; (e) $R^2 = 0.9656$. The values of R^2 parameter for anodic process (a) $R^2 = 0.9950$; (b) $R^2 = 0.9891$; (c) $R^2 = 0.9913$; (d) $R^2 = 0.9781$; (e) $R^2 = 0.9243$.

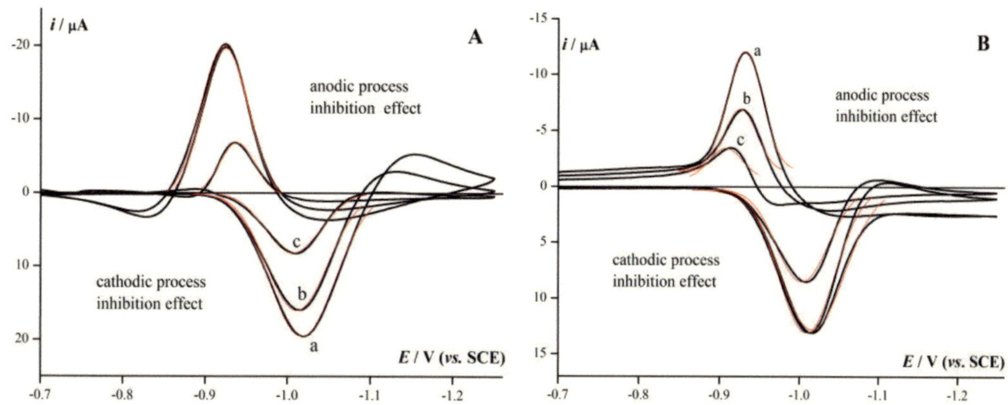

Figure 4. A comparison of experimental electroinhibition effects of CPR type, determined from eq. (2) for different concentrations (a) 0.1 M, (b) 0.05 M, (c) 0.01 M of *α*-cyclodextrin (A) and *β*-cyclodextrin (B). The parent electrode process: CV electroreduction of 1 mM Zn^{2+} in 1M $NaClO_4$ on Hg. Scan rate 1.003 V s^{-1}. (A) The Gaussian fit (red curves) results in good correlation parameters, for example the values of R^2 parameter of Gaussian fit for cathodic process are: (a) $R^2 = 0.9956$; (b) $R^2 = 0.9920$; (c) $R^2 = 0.9946$. The values of R^2 parameter for anodic process are: (a) $R^2 = 0.9972$; (b) $R^2 = 0.9995$; (c) $R^2 = 0.9905$. (B) The Gaussian fit (red curves) results in the good correlation parameters for example the values of R^2 parameter of Gaussian fit for cathodic process are: (a) $R^2 = 0.9936$; (b) $R^2 = 0.9914$; (c) $R^2 = 0.9804$. The values of R^2 parameter for anodic process are: (a) $R^2 = 0.9861$; (b) $R^2 = 0.9479$; (c) $R^2 = 0.8379$.

The results indicate that the influence of cyclodextrins on the electroreduction process is not uniform and that *α*–CD is effective as an inhibitor, which remains in agreement with literature data [12]. The data presented in Fig. 3 for alcohols and in Fig. 4 for cyclodextrins have shown, that it is possible to obtain the picture of inhibitory effect more pronounced than in case of applying the single value of kinetic parameter e.g. k_{app} or k_1, k_2 alone. The Gaussian fit is also included for data in Fig. 3 – 4. It is evident that fully shaped curves are obtained for relatively high concentration of the adsorbate. The quality of Gaussian fit decreases with the lowering of the adsorbate concentration as it could have been expected.

The detailed investigations with the use of cyclodextrins as adsorbates have shown that the adsorption process is complex with the possibility of compact layers or multi-layers, host-guest complexes formation [13]. Our results with good quality Gaussian fit suggest that both complex (α, β-cyclodextrins) and simple adsorption (alcohols) undergo Gaussian distribution.

Application of CV method usually leads to a series of peaks corresponding to the different scan rates and so it is in our case. However, CV currents recorded at different scan rates are not comparable and should be normalized by $v^{1/2}$ factor. The different scan rate data, performed in such a way, are presented in form of *CPR* curves in Fig. 5. In other words, instead of parent CV peaks for regular and inhibited process, the final *CPR* curves can be taken under consideration.

On the other hand, the $Q = f(v)$ plot (Fig. 6A) makes it possible to express the inhibition effect quantitatively and normalization by $v^{1/2}$ factor is not needed. In turn, the dependence of the relative inhibition effect $(Q_{CD} - Q_{Zn})/Q_{Zn}$ (where Q_{Zn}, Q_{CD} are charges obtained from integrated cathodic parts of CV curves for electroreduction of Zn^{2+} 1 mM in 1 M $NaClO_4$ in absence and presence of α,β-CD, respectively) at different concentrations and scan rates is presented in Fig. 6B.

Apart from better visualization and separation of adsorption effect from "mixed" current response, CPR approach gives the possibility of adsorption isotherm determination. A respective attempt is presented in Fig. 6B. An assumption was made that a charge Q_{CD} determined from CPR curve is proportional to the amount of adsorbed species. The similar assumption is commonly accepted when adsorption phenomena are investigated by differential capatitance method. A change ∂C is assumed to be proportional to ∂Q and to Γ. The results of Langmuir fit indicate that the obtained parameters are reasonable in the range of the published data [41].

The coincidence with the Gaussian function is expected not to be specific for studied system Zn^{2+}/alcohols and Zn^{2+}/cyclodextrin. It should be observed for at least other qusireversible-irreversible reduction/inert organic adsorbate systems in which pre-peaks or post-peaks do not appear.

It is very probable that numerous factual current responses (CR) and not only curves affected by the effects of inhibition, lead to similar CPR curves with almost identical shapes. Consequently, any change in electrochemical response caused by the influence of added adsorbate, complexing agent or other factors is manifested by, at least, the following influences: positive/negative shift of CR along potential axis and/or change of the CR slope and/or decrease/increase of its high. Above effects occur together, but usually one of them dominates. Therefore, the observed CPR effect, that is an output function, must be interpreted in frames of its parent experimental input.

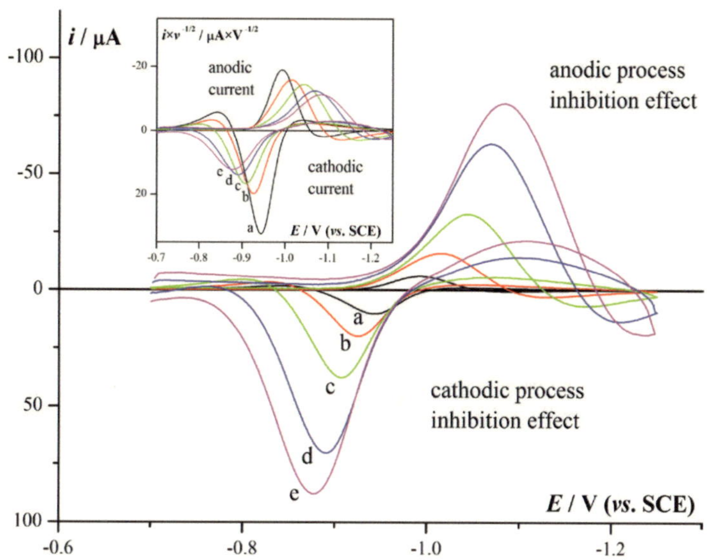

Figure 5. A comparison of experimental electroinhibition effect of CPR type for 1 mM Zn^{2+}, 1 M $NaClO_4$, 0.05 M α-CD on Hg electrode at different scan rate in V s^{-1}: (a) 0.1, (b) 1.003, (c) 5.12, (d) 25.6, (e) 51.2. Gaussian fits for a,b,c,d,e plots, not shown on the plot, are of very good quality by statistics. The normalized ($i \times v^{-1/2}$) currents of CPR curves are shown in the additional sub-window. The parent CV responses are not shown.

Another experimental examples of CPR effects by CV and NPV techniques are presented in Fig. 7 and Fig. 8 for various adsorbates and various electrode processes.

Data of Fig. 8 A and B present the influence of added organic adsorbates as thiourea (part A, A') and *N,N*-dimethylaniline (part B, B') on molecular oxygen electroreduction in DMF and water, respectively. Fig. 9 presents the case in which the influence of substituent on adsorbate molecule on fixed electrode process is visible. The catalytic effect increases once the electron density on nitrogen atom in thiourea increases. The effect on substituent on thiourea molecule on apparent rate constant is described in paper by Ikeda and co-authors [45].

The results indicate that CPR effect is observed also on GCE and in nonaqueous medium (Fig. 8A).

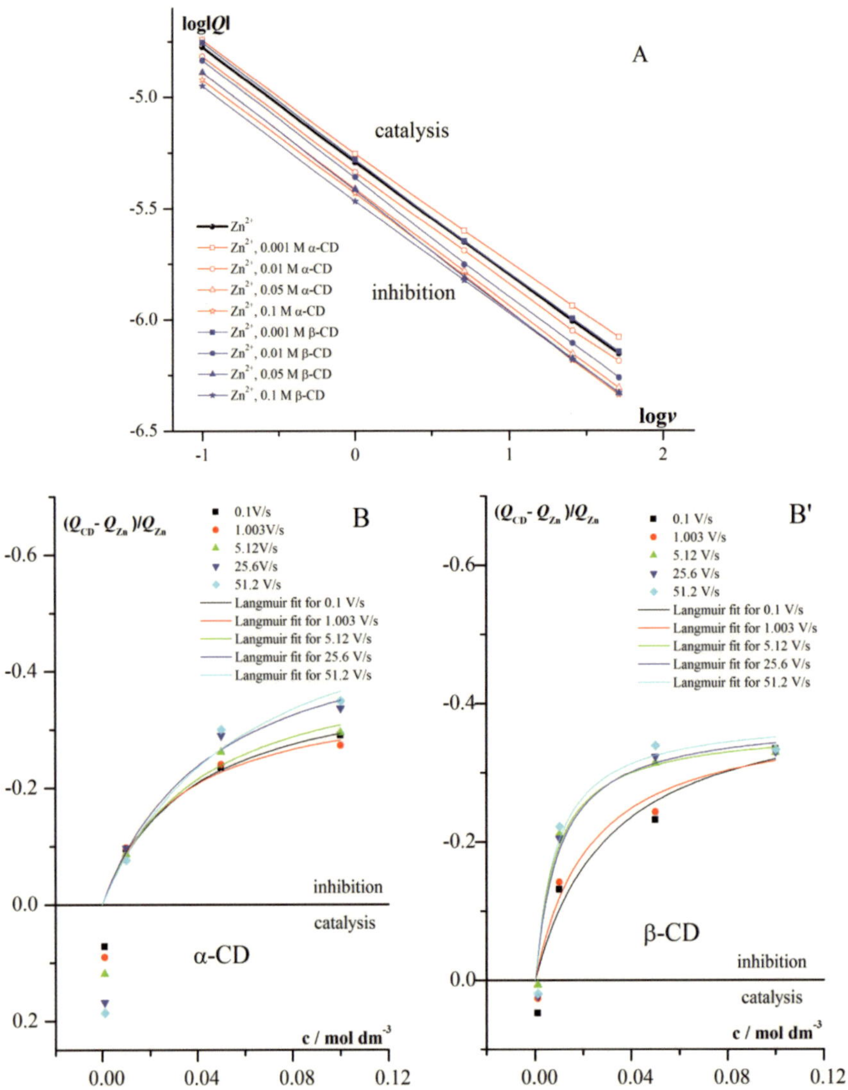

Figure 6. (A) The logarithmic dependence of the charge obtained by integration of cathodic scans of experimental CV curves for 1 mM Zn^{2+}electroreduction in 1.0 M $NaClO_4$ on Hg electrode in absence and presence of *α*-CD and *β*-CD against scan rate. Note that for low 0.001 M concentration of *α*-CD and *β*-CD the inhibition effect changes into the catalytic effect (two lines above thick black line for Zn^{2+} reduction alone refer to catalysis). (B) The normalized CPR effect presented in form of adsorption isotherm obtained on the basis of respective integrated current responses. The scatter plots represent the dependence of the relative inhibition effect for CV electroreduction of Zn^{2+} 1 mM in 1M $NaClO_4$ on Hg electrode in function of *α*-CD and *β*-CD concentration at different scan rates. Q_{Zn}, Q_{CD} – charges obtained from integrated cathodic parts of CV curves. The curves represent the Langmuir fit in form of eq.: $q = q_s Kc/(1+Kc)$, where q, q_s – surface excess, surface excess at saturation and K – equilibrium constant of adsorbate at surface. The examples of values of parameters of Langmuir fit for *α*-CD: at v = 0.1 V s^{-1}; $q_s = -0.40 \pm 0.19$, $K = 27 \pm 33$, $\chi^2 = 3.5\times10^{-3}$, $R^2 = 0.911$; and for *β*-CD: at 51.2 V s^{-1}; $q_s = -0.38 \pm 0.05$, $K = 113 \pm 71$, $\chi^2 = 2.2\times10^{-3}$, $R^2 = 0.949$. Note that for low concentrations of *α*-CD and *β*-CD inhibition effect changes into catalytic effect. The effect shape of plot resembles the Langmuir plot.

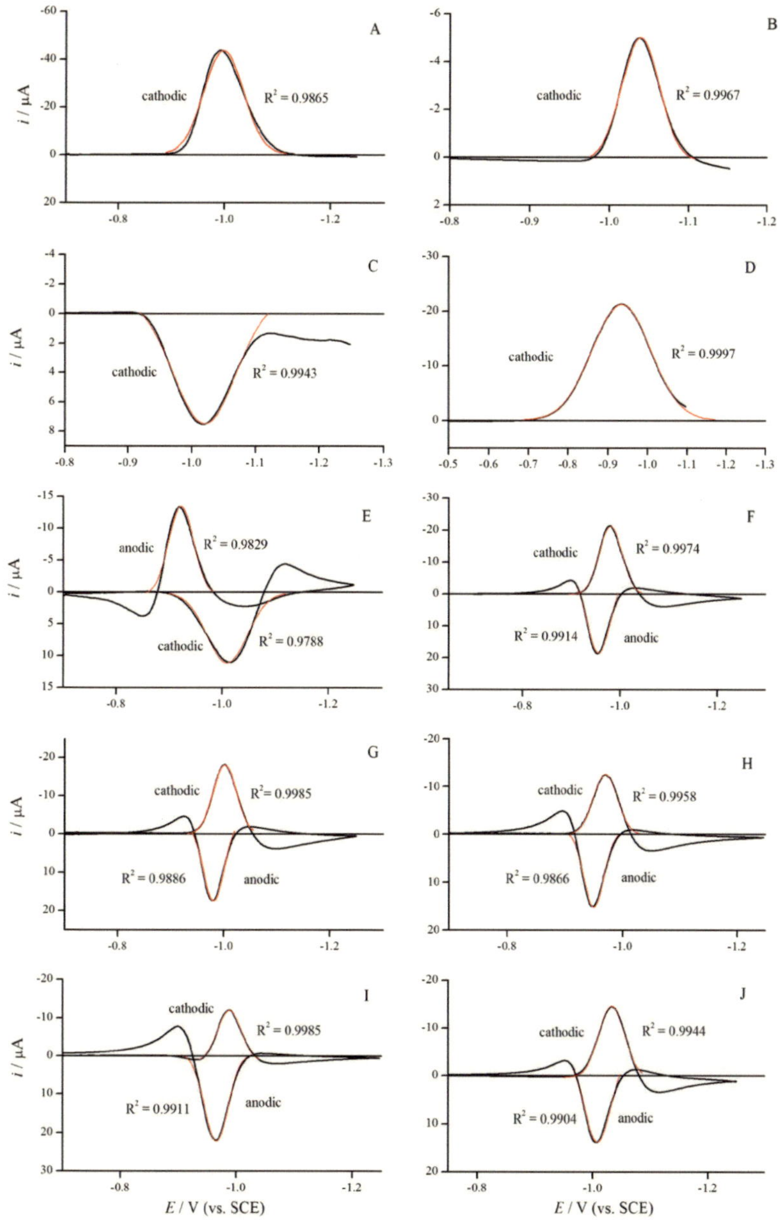

Figure 7. A–J. Examples of CPR results (parent responses are not shown) for different electrode process/adsorbate systems obtained with the use of CV and NPV techniques. (Plots A-C): The NPV electroreduction of 1 mM Zn^{2+} in 1 M $NaClO_4$ in H_2O on Hg electrode in presence: (A) – 5 mM *N,N'*-dimethylthiourea; (B) – 5 mM *N,N,N',N'*-tetramethylthiourea; (C) – 2 μM dextran. (Plot D): The NPV electroreduction of molecular oxygen about 4 mM in 0.3 M TBAP in DMF on GCE3 in presence 1 mM *N,N*-dimethylaniline. (Plots E-J): The CV electroreduction of Zn^{2+} in 1 M $NaClO_4$ in H_2O on Hg electrode in presence of: (E) – 10 μM dextran; (F) – 2 mM 3,4-diaminotoluene; (G) – 55 mM thiourea; (H) – 5 mM *N,N'*-dimethylthiourea; (I) – 5 mM *N,N'*-diethylthiourea; (J) – 5 mM N,N,N',N'-tetramethylthiourea. Scan rate 1 V s^{-1}. The Gaussian fit (red curves) and the values of R^2 correlation parameter are shown on the plot.

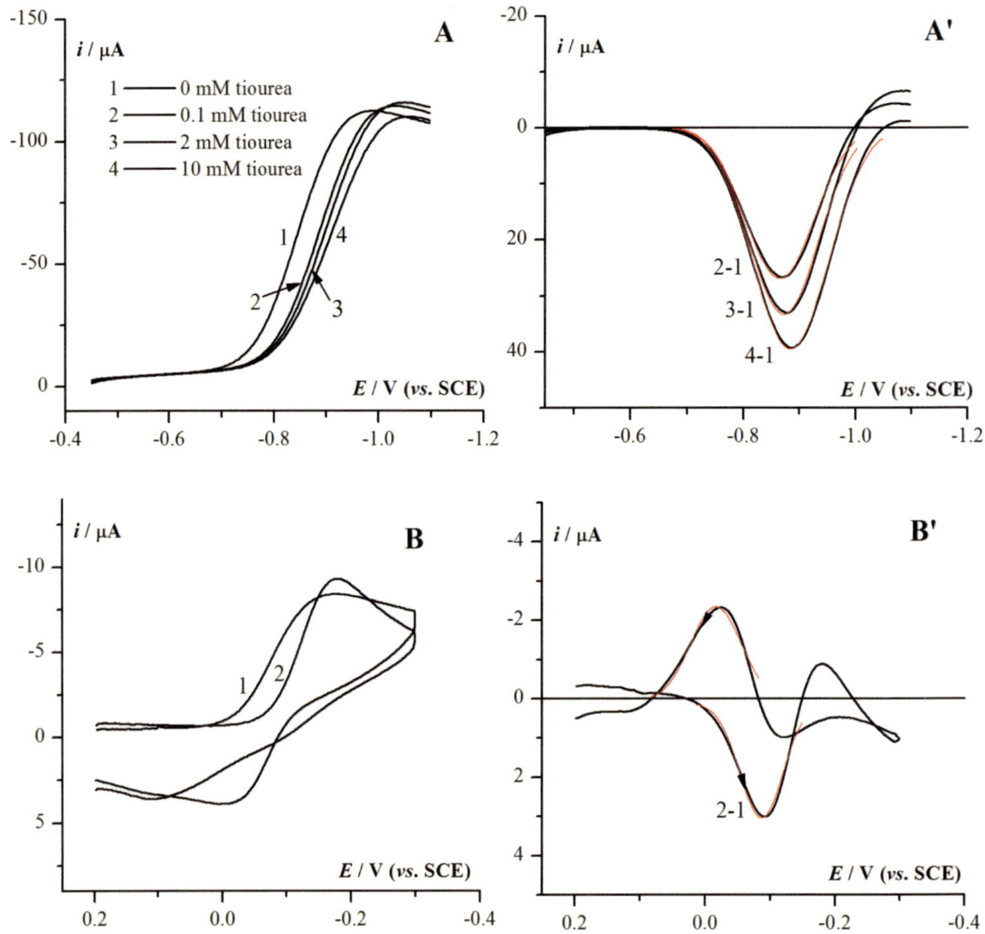

Figure 8. Further experimental examples of CPR inhibition effect. (A) The effect of thiourea on one-electron normal pulse voltammetry reduction of molecular oxygen in 0.3 M TBAP in DMF on GCE3. Thiourea concentration 0, 0.1, 2, 10 mM, O_2 concentration about 4 mM. (A) The same data in CPR form. The Gaussian fit (red curves) leads to good correlation parameters. For example the values of R^2 parameter of Gaussian fit are: (2–1) $R^2 = 0.9926$; (3–1) $R^2 = 0.9913$; (4–1) $R^2 = 0.9920$. (B) The effect of *N,N*-dimetyloaniline on two-electron electroreduction of molecular oxygen by CV in 0.2 M KNO_3 in H_2O on Hg electrode. Scan rate 1 V s^{-1}, *N,N*-dimetyloaniline concentration 1mM, O_2 concentration about 1 mM. (B') The same data in CPR form. The Gaussian fit (red curves) results in the good correlation parameters. For example the values of R^2 parameter of Gaussian fit are 0.9738 and 0.9628 for cathodic and anodic process, respectively.

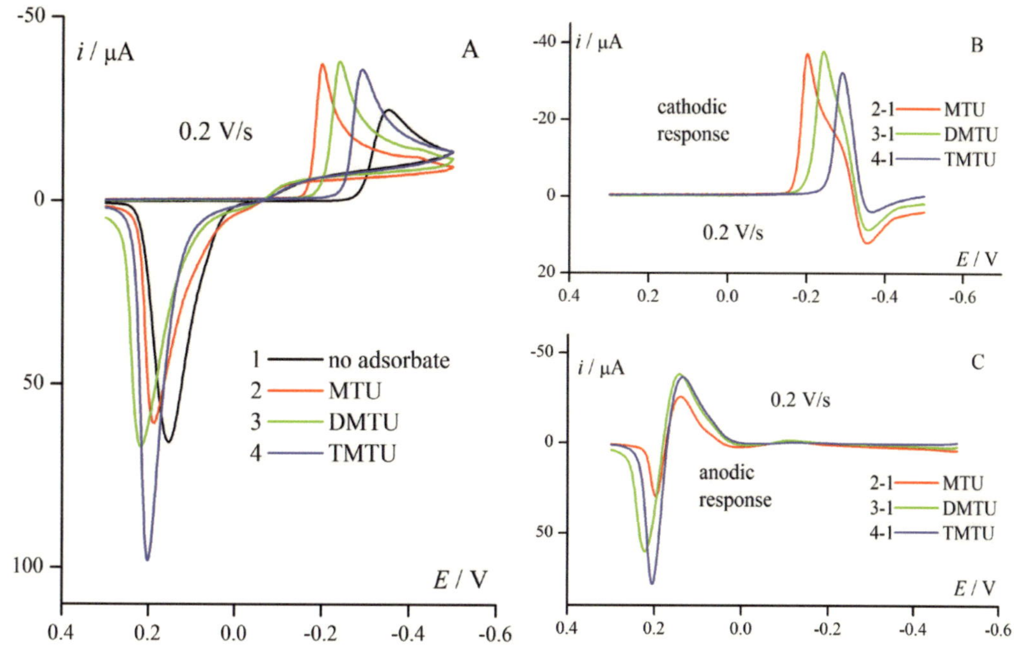

Figure 9. Further examples of CPR effect. Electroreduction of 1 mM Bi^{3+} in 0,1 M $HClO_4$ on GCE2 in presence and absence of 10 mM substituted thioureas (N-methylthiourea (MTU), N,N'-dimethylthiourea (DMTU), N,N,N',N'-tetramethylthiourea (TMTU)). The increase of the catalytic substituent effect from MTU through DMTU to TMTU is visible on position and height of cathodic responses. Scan rate 0.2 V s^{-1}. The parent CV responses are not shown.

4.2. CPR effect as a result of theoretical model

Since the presented results are only an experimental and fenomenological view of the problem, the respective theoretical background is needed. It turned out that the Gaussian-shaped curves of *CPR* type can be obtained also theoretically by numerical simulation. We used the theoretical kinetic EE ‖ Hg(Zn) model and *ESTYM_PDE* software, both described in *Kinetics* section. The literature values of kinetic parameters determined for Zn^{2+} electroreduction in the 1 M sodium perchlorate solution [18] were applied. The EE mechanism provides the simple but reliable model among others proposed for Zn^{2+} electroreduction [38,39,46-49]. The results of simulation are presented in Fig. 10. It turned out that increase/decrease of electrochemical rate constant corresponds to a change of CV curve which is identical to Gaussian *CPR* experimental curve. For clarity, the results of Gaussian fit are shown separately in Fig. 10B. The obtained results of performed simulation provide an indirect evidence of assumption commonly applied in literature that electrocatalysis/electroinhibition effect can be described by increase/decrease of heterogeneous rate constant [15]. The simulations confirm the literature adsorption model in which the effective observed current is the superposition of both currents on the free surface and on covered fraction i.e. "adsorption" current. It is assumed that the rate constants on covered and uncovered surface should be different (Fig. 10 – simulation). Moreover, the resolution of the experimental currents for two reductions gives possibility for calculation of individual rate constants. It has to be noticed, however, that the mathematical analysis described in reference [50] indicates that the effects of inhibition can be much more complex than a simple k_s decrease analysis presented in Fig. 10.

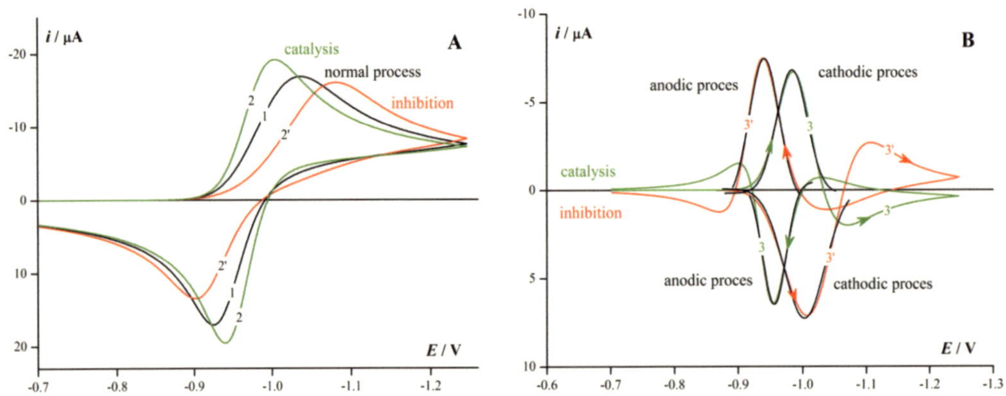

Figure 10. (A) An example of CV simulation results obtained for EE ‖ Hg(Zn) mechanism (section *Kinetics* eq.(1)) with the following kinetic parameters: $v = 1.0$ V s^{-1}, $D_1 = D_2 = 4.0\times10^{-6}$ cm^{2}s^{-1}, $D_3 = 1.67\times10^{-5}$ cm^{2}s^{-1}, $E_{0,1} = -1.049$ V $E_{0,2} = -0.904$ V, $A = 0.0159$ cm^2 (spherical electrode), $\alpha_1 = \alpha_2 = 0.67$; (1) $k_1 = k_{-1} = 0.031$, $k_2 = k_{-2} = 0.35$; (2) $k_1 = k_{-1} = 0.09$, $k_2 = k_{-2} = 0.9$; (2') $k_1 = k_{-1} = 0.01$, $k_2 = k_{-2} = 0.1$ (rate constants in cmxs^{-1}); the values of kinetic parameters were taken from [18]. Initial concentration of substrate was $c_0 = 1$ mM. Curves 1, 2 and 2' were obtained from the simulated CV curves for normal process, electrocatalysis and electroinhibition, respectively. (B) Curves 3 (green) and 3' (red) represent the *Gaussian current response* and were obtained from the simulated CV curves (A) for electrocatalysis and electroinhibition effect, respectively. The Gaussian fits (black curves) result in good correlation parameters. For example the values of R^2 parameter for curve 3 for cathodic and anodic process are 0.9776 and 0.9848 respectively; for curve 3' for cathodic and anodic process R^2 values are 0.9770 and 0.9896, respectively.

Despite the fact that the results of modeling of the CPR effect have been presented (Fig. 10), this paper is based mainly on experimental facts. Deeper theoretical analysis of observed phenomena is not possible at the present state of knowledge.

4.3. The CPR effect obtained from processed literature data

In order to extend the scope of experimental data concerning CPR effect, the CPR effect obtained from processed literature data [5,14,51] is presented in Figs. 11-14. The influence of added organic adsorbate, namely polyvinyl alcohol, on VO^{2+} reduction is shown in Fig. 11 A and A' (source of data: [5b]).

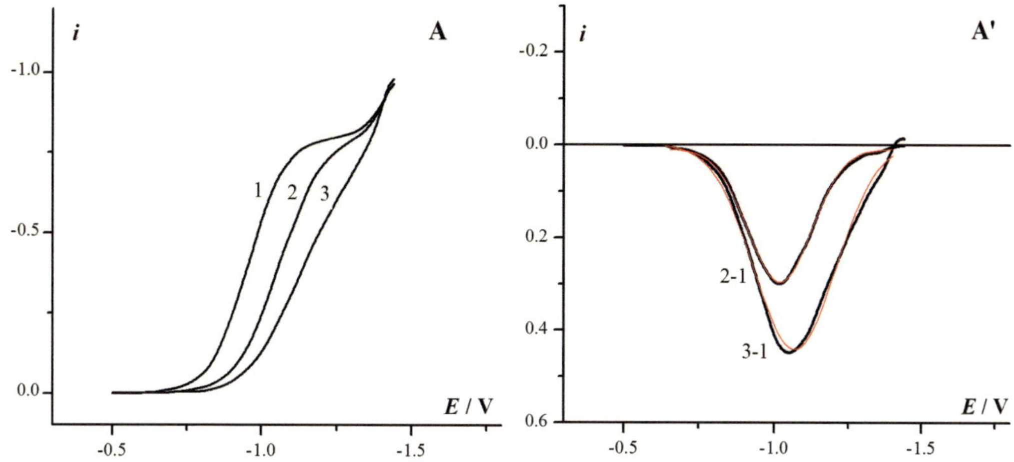

Figure 11. Further experimental examples of CPR inhibition effect derived by digitalization and processing of literature data. (A) The effect of polyvinyl alcohol on the reduction of VO^{2+}. Curve (1) – 3 mM $VOSO_4$, 0.1 M H_2SO_4. Polyvinyl alcohol concentration: (2) – 0.005%, (3) – 0.0075%. Source of data [5b]. (A') The same data in CPR form. The Gaussian fit (red curves) leads to good correlation parameters. For example the values of R^2 parameter of Gaussian fit are: (2–1) $R^2 = 0.9988$; (3–1) $R^2 = 0.9930$.

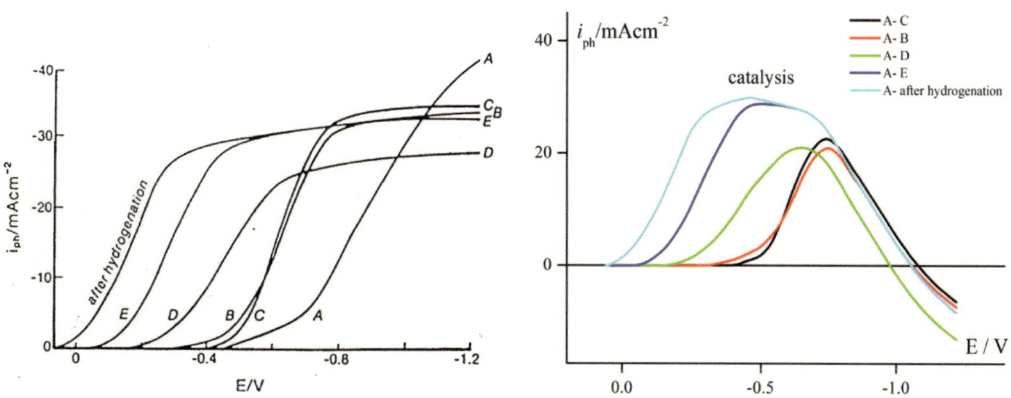

Figure 12. Examples of CPR effect (right plot) derived by digitalization and processing of literature data (left plot) concerning photocurrent-potential relation for p-Si electrode (Fig. 1 p. 206, [51], with kind permission from Springer Science+Business Media B.V.). The similar picture (not shown) can be obtained basing on Fig. 4, p. 208, in monograph [51]. Symbols ABCDE denote the surface etching procedure described therein.

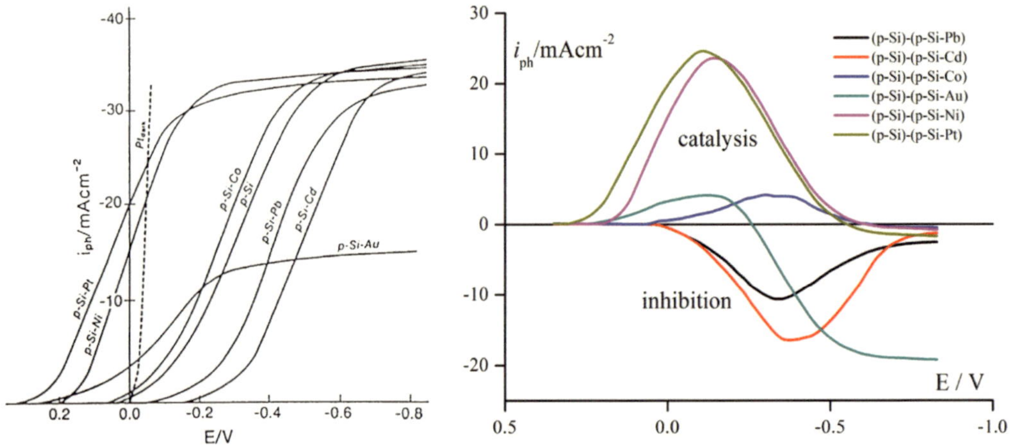

Figure 13. Examples of CPR effect (right plot) derived by digitalization of literature data (left plot) covering potentiodynamic runs recorded for p-Si and p-Si-Me electrodes (Fig. 3 p. 207, [51], with kind permission from Springer Science +Business Media B.V.). The similar picture (not shown) can be obtained basing on Fig. 5, p. 208, in monograph [51].

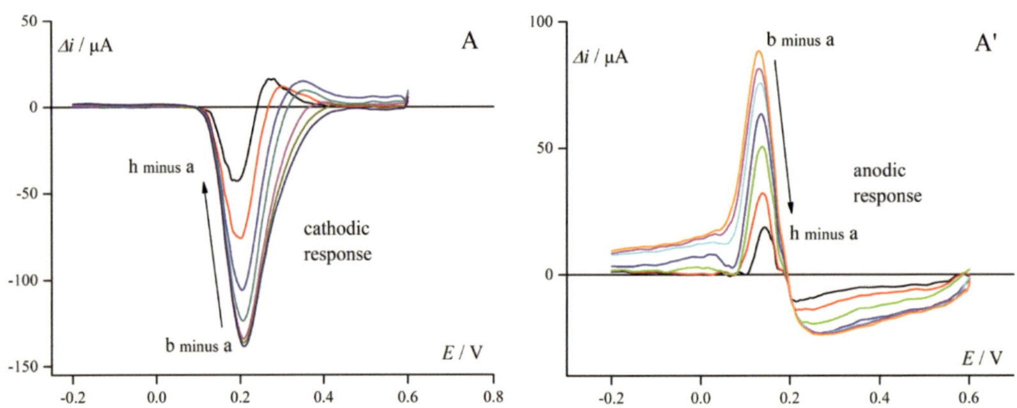

Figure 14. Another examples of CPR effect derived by digitalization and processing of literature CV data [14] (with kind permission from Wiley) for electroreduction of 1.0 mM 4-*tert*-butylcatechol in the presence of various concentrations of β-CD at pH 7.0. Concentration of β-CD from (a) to (h): 0.0, 0.3, 0.5, 1.0, 2.0, 4.0, 7.0, and 10.0 mM, respectively. Scan rate: 2 V s^{-1}.

Stochastic nature of adsorption phenomena is also visible on surface tension π vs. potential E dependences presented in monograph [15,16]. The *CPR* peaks involved in adsorption phenomena are similar to ones obtained by means of chromatography. The latter also undergo Gaussian statistics [52]. Both facts confirm the probabilistic character of chemical dynamics [53].

4.4. CPR effects for auto-electrocatalysis processes (ACPR)

The hitherto presented data respected to the electroreduction of inorganic cations in the presence of non-electroactive organic adsorbate. Some experimental facts indicate that the idea of CPR can be extended on electroactive adsorbates. In paper [17] we have detected the autocatalytic effect caused by organic electroactive substance i.e. that which playes double role of organic adsorbate and electroactive substance. The autocatalytic effect was revealed by comparison of experimental and mathematical model responses as well as by comparison of results obtained on Hg (autocatalytic effect present) and GCE electrode (no autocatalytic effect).

In order to show more quantitative explanation of the influence of the autocatalytic effect on the enhancement the cathodic current observed on Hg electrode the difference between experimental and model curves will be shown in Fig. 15.

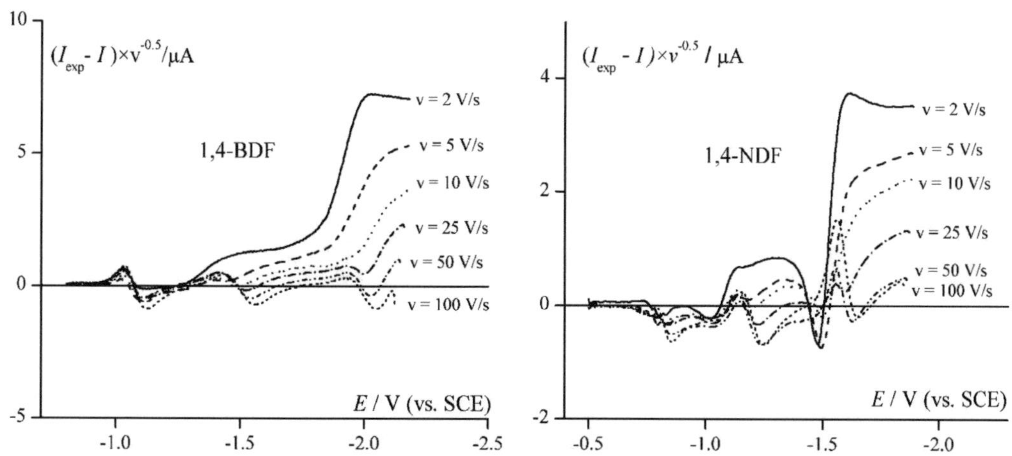

Figure 15. The "difference curves" (normalized vs. scan rate) between experimental (affected by adsorption) and model responses of the CV electroreduction of 1,4-BDF and 1,4-NDF in 0.3 M TBAP in DMF. The result can be considered as an example of auto-electrocatalytic effect. No smoothing of the data was applied. [Reprinted from *Electroanalysis*, Vol. 18, Sanecki, P., Skitał, P., Kaczmarski, K., Numerical Modeling of ECE – ECE and Parallel EE – EE Mechanisms in..., 981-991, Copyright (2006), with permission from Wiley].

From the character of the difference curves of CPR kind one may conclude that observed effect is not random but rather systematic with the wave profile increasing with the decreased scan rate. The observed autocatalytic effect vanish at high scan rates, giving discrepancy low enough to permit the usage of these data for the estimation of kinetic parameters [17]. If the effect is connected with adsorption of electrode reaction product, the adsorption process would be slow in comparison to the time window of the high scan rate experiment. Similar observations for inorganic reactant-organic substance/halide ion systems have been written down by the others, by whom catalytic effect was explained by *bridging model* [49] or *surface reaction model* (see [54] and the literature cited therein).

Interesting electrocatalysis/electroinhibition examples of CPR type effects can be provided by early published polarographic data. Maxima of first kind (sharp) and second kind (diffused) appearing on diffusion/convection controlled limiting currents generally are explained by adsorption of electroactive substance and/or a movement of the solution near the electrode surface. Polarographic maxima were recognized as a case of electrocatalysis [5c, 55]. Since maxima hinder the precise evaluation of polarographic curves, they (maxima) are removed (damped) by introduction of strong surface active compounds as high-molecular organic compounds e.g. dyes (fuchsin). It means that a adsorption competition between electroactive adsorbate and nonelectroactive adsorbate takes place.

The adsorption of electroactive substance can be a source of its additional amount on electrode surface and consequently of limiting current excitation. Fig. 16 presents an example of such kind of auto-electrocatalysis, in which asymmetric curve, of autocatalytic CPR kind (ACPR), characteristic to first kind maximum, appears (source of data [55]).

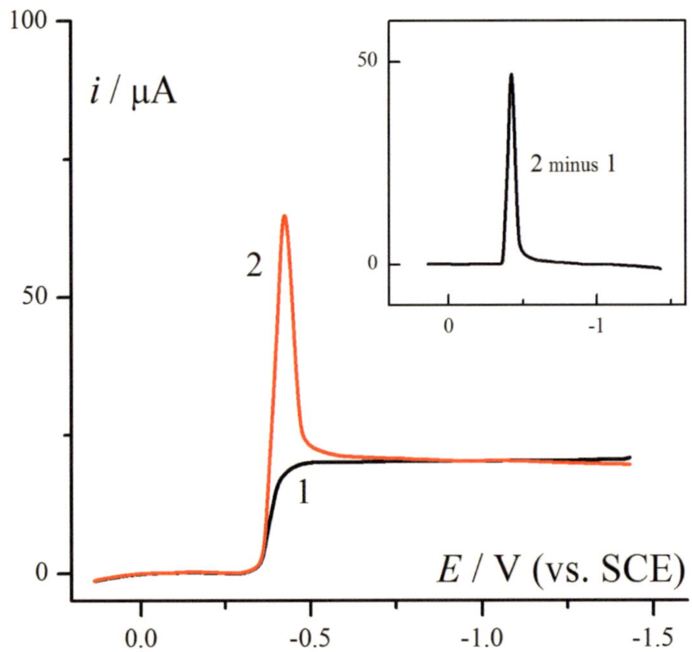

Figure 16. An example of electrocatalysis caused by adsorption of electroactive substance i.e. Pb^{2+} ions. The process can be considered as autocatalytic CPR effect (ACPR). The polarographic reduction of 2.3 mM Pb^{2+} in 0.1 M KCl solution: (curve 1) in absence of maximum suppressor; (curve 2) after the addition of 0.0002% sodium methyl red; inset window (curve 2 minus 1) i.e. affected curve minus regular curve. Curves 1 and 2 were obtained from processed data of Figure 6.8, p. 317 in [55].

The symmetric difference curve of ACPR kind can be obtained from second kind maxima data Fig. 17 ([5] Fig. XVIII-8, p.419).

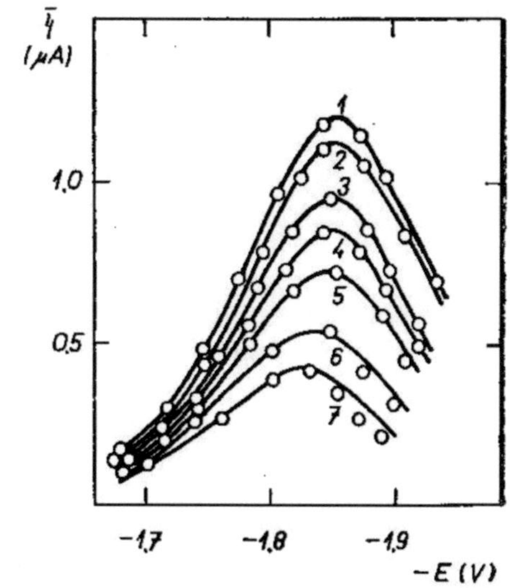

Figure 17. Another example of autocatalytic ACPR effect obtained from literature data [5]. The influence of ionic strength on catalytic hydrogen evolution waves in 3 mM solution of quinine in borate buffer pH 9.5. Concentrations of Na^+: (1) 0.04 M; (2) 0.045 M; (3) 0.05 M; (4) 0.055 M; (5) 0.06 M; (6) 0.07 M; (7) 0.08 M. [Reprinted from [5] Heyrovský, J.; Kůta, J. Principles of Polarography, Publishing House of the Czechoslovak Academy of Sciences: Prague 1965, Fig. XVIII-8, p.419].

The gradual passing from first kind maximum (a) to second kind maximum (f), presented in [5], can be interpreted as a change of adsorption mechanism visible as the change of shape of probabilistic curve (Fig. 18).

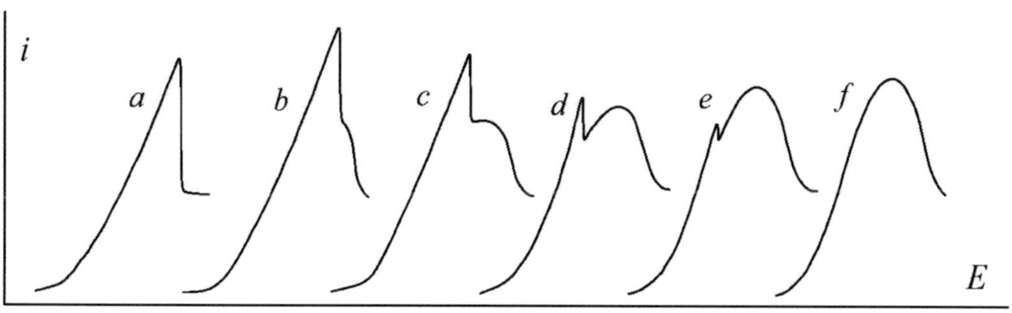

Figure 18. Another example of electrocatalysis of CPR type caused by adsorption of electroactive substance namely i.e. molecular oxygen. Solution 0.01 M KCl, exposed to air. The change of electrochemical response was caused by mercury head height lowering (processed literature data from Heyrovský, J.; Kůta, J. Principles of Polarography, Publishing House of the Czechoslovak Academy of Sciences: Prague 1965, Fig. XIX-23, p.457 [5].

On the other side, electroinhibition examples can also be recognized as characteristic minima on limiting currents of anions electroreduction e.g. $S_2O_8^{2-}$ (Figures XIV-17, XIV-21, XIV-22 in [5]). They also yield the difference curves of the symmetric probabilistic shape. Their precise explanation is much more complex and should be considered in frames of double layer structure changes.

5. Conclusions

1. The faradaic electrode process of electroreduction of inorganic cations creates a reference system on which the adsorption / inhibition effect can be observed.
2. The experimental catalysis/inhibition effect of electrode process affected by adsorption can be quantitatively separated from complex current response and displayed in the form of probabilistic type faradaic currents. The CPR curves can be considered as a quantitative adsorption characteristic of an electroinactive organic adsorbate. In many cases, the inhibition and catalysis processes are mutually coupled.
3. The results obtained by numerical simulation for EE process indicate that CPR effect is not accidental and that the assumption, that organic substance increases or decreases the apparent or individual rate constant of the electrode process, which is commonly accepted in literature is of value.
4. In the CPR presentation the effect of adsorption is separated and defined as well as can be processed and discussed quantitatively e.g. as a function of scan rate, a concentration of species in solution, etc. The shape of CPR curve (symmetrical or not symmetrical) may indicate the presence of possible interactions between molecules of adsorbate (Langumir or Frumkin model of isotherm).
5. The CPR effect is also observed for systems in which an organic or inorganic adsorbate is an electrode active reagent simultaneously (autocatalytic ACPR case).
6. Electrocatalysis/electroinhibition effect is a positive/negative difference between electrochemical response after and before introduction of a non-electroactive substance into solution and/or on electrode surface or a positive/negative difference between electrochemical response affected and non affected by substrate/product adsorption. The difference (eq. (2)) can be expressed quantitatively in the bell-shaped peak form of *Current Probabilistic Response*.

Abbreviations, symbols and acronyms

CPR – probabilistic type current response ("difference curves") obtained from eq. (2)

ACPR – autocatalytic CPR effect

CV – cyclic voltammetry

NPP – normal pulse polarography

EE – set of two consecutive one electron steps

α-CD – α-cyclodextrin

β-CD – β-cyclodextrin

SCE – saturated calomel electrode

TBAP – tetrabuthyl ammonium perchlorate

DMF – dimethylformamide

Q – charge

1,4-BDF – 1,4-benzenedisulfonyldifluoride

1,4-NDF – 1,4-naphthalenedisulfonyldifluoride

Tables

Adsorbate	Electrochemical. reaction, Electrode material	Figure	Observed effect	Literature
N,N'-dimethylthiourea, 3,4-diaminotoluene	$Zn^{2+} + 2e^- = Zn$, Hg	Fig. 1	CPR	this paper
n-pentanol		Fig. 2A, B	CPR	this paper
n-hexanol, *n*-pentanol, *t*-pentanol, *n*-butanol, *n*-propanol		Fig. 3A, B	CPR	this paper
α-cyclodextrin		Fig. 4A	CPR	this paper
β-cyclodextrin		Fig. 4B	CPR	this paper
α-cyclodextrin		Fig. 5	CPR	this paper
N,N'-dimethylthiourea		Fig. 7A	CPR	this paper
N,N,N',N'-tetramethylthiourea		Fig. 7B	CPR	this paper
dextran		Fig. 7C	CPR	this paper
N,N-dimethylaniline	$O_2 + e^- = O_2^{-\cdot}$, GCE	Fig. 7D	CPR	this paper
dextran	$Zn^{2+} + 2e^- = Zn$, Hg	Fig. 7E	CPR	this paper
3,4-diaminotoluene		Fig. 7F	CPR	this paper
thiourea		Fig. 7G	CPR	this paper
N,N'-dimethylthiourea		Fig. 7H	CPR	this paper
N,N'-diethylthiourea		Fig. 7I	CPR	this paper
N,N,N',N'-tetramethylthiourea		Fig. 7J	CPR	this paper

Adsorbate	Electrochemical. reaction, Electrode material	Figure	Observed effect	Literature
thiourea	$O_2 + e^- = O_2^{-\cdot}$, GCE	Fig. 8A	CPR	this paper
N,N-dimetyloaniline	$O_2 + 2e^- + 2H^+ = H_2O_2$, Hg	Fig. 8B	CPR	this paper
N-methylthiourea, N,N'-dimethylthiourea, N,N,N',N'- tetramethylthiourea	$Bi^{3+} + 3e^- = Bi$, GCE	Fig. 9	CPR	this paper
modeling via *k* variation	–	Fig. 10	CPR as a model effect	this paper
polyvinyl alcohol	$V^{4+} + e^- = V^{3+}$, Hg	Fig. 11	CPR	[5b]
Modification of electrode surface	photocurrent, p-Si electrode	Fig. 12	CPR	[51]
Modification of electrode surface	photocurrent, p-Si and p-Si-Me electrodes	Fig. 13	CPR	[51]
β-cyclodextrin	Electroxidation of 4-*tert*-butylcatechol (H_2Q) $H_2Q - 2e^- = Q + 2H^+$, GCE	Fig. 14	CPR	[14]
1,4-BDF 1,4-NDF	$ArSO_2F + 2e^- = ArSO_2^- + F^-$, GCE	Fig. 15	ACPR	[17]
Pb^{2+} ions	$Pb^{2+} + 2e = Pb$, Hg	Fig. 16	ACPR	[55]
quinone	$H^+ + e = H$	Fig. 17	ACPR	[5]
O_2	O_2/H_2O_2, Hg	Fig. 18	ACPR	[5]

Table 1. The list of adsorbates and the electrode processes applied in the present study.

Author details

Piotr M. Skitał and Przemysław T. Sanecki*

*Address all correspondence to: psanecki@prz.edu.pl

Faculty of Chemistry, Rzeszów University of Technology, Rzeszów, Poland

This chapter is dedicated to professor Zbigniew Galus.

References

[1] Krjukova AA, Loshkarev MA. O prirode tormozyashchego deistviya poverkhnostno-aktivnykh-veshchestv na elektrodnye protsessy. Zh. Fiz. Khim. 1956;30: 2236–2243.

[2] Schmid R, Reilley CN. Concerning the effect of surface-active substances on polarographic currents. J. Am. Chem. Soc. 1958;80: 2087–2094.

[3] Weber J, Koutecký J, Koryta J. Ein Beitrag zur Theorie der polarographischen Ströme, die durch Adsorption eines elektroinaktiven Stoffes beeinflußt sind. Z. Elektrochem. 1959;63: 583–588.

[4] Weber J, Koutecký J. Theorie der durch die Adsorption des elektroinaktiven Stoffes bei einer reversiblen Elektrodenreaktion beeinflussten polarographischen Ströme. Collect. Czech. Chem. Commun. 1960;25: 1423–1426.

[5] Heyrovský J, Kůta J. Principles of Polarography. Prague: Publishing House of the Czechoslovak Academy of Sciences; 1965. a) p309. b) p315. c) p459.

[6] Sykut K, Dalmata G, Nowicka B, Saba J. Acceleration of electrode processes by organic compounds - "cap-pair" effect. J. Electroanal. Chem. 1978;90: 299–302.

[7] Saba J, Nieszporek J, Gugała D, Sieńko D, Szaran J. Influence of the mixed adsorption layer of 1-butanol/toluidine isomers on the two step electroreduction of Zn(II) ions. Electroanalysis 2003;15: 33–39.

[8] Marczewska B. Mechanism of the acceleration effect of thiourea on the electrochemical reduction of Zinc(II) ions in binary mixtures on mercury electrode. Electroanalysis 1998;10: 50–53.

[9] Sanecki P. A distinguishing of adsorption-catalyzed and regular part of faradaic current for inorganic cation-organic adsorbate system: propabilistic curves in cyclic voltammetry and normal pulse polarography. Electrochem. Comm. 2004;6: 753–756.

[10] Golędzinowski M, Kisova L, Lipkowski J, Galus Z. Manifestation of steric factors in electrode kinetics. Investigations of the deposition and dissolution kinetics of the Cd^{2+} /Cd(Hg) system in presence of adsorbed aliphatic alcohols and acids. J. Electroanal. Chem. 1979;95: 43–57.

[11] Jaworski RK, Golędzinowski M, Galus Z. Adsorption of α-, β- and γ-cyclodextrins on mercury electrodes from 1M $NaClO_4$ and 0.5M Na_2SO_4 aqueous solutions. J. Electroanal. Chem. 1988;252: 425–440.

[12] Golędzinowski M. The influence of α-, β- and γ-cyclodextrins on the kinetics of the electrode reactions in 1M $NaClO_4$ and 0.5M Na_2SO_4 aqueous solution. J. Electroanal. Chem. 1989;267: 171–189.

[13] Hromadova M, de Levie R. A sodium-specific condensed film of α-cyclodextrim at the mercury / water interface. J. Electroanal. Chem. 1999;465: 51–62.

[14] Rafiee M. Electrochemical Oxidation of 4-tert-Butylcatechol in the Presence of β-Cyclodextrin: Interplay between E and CE Mechanisms. Int. J. Chem. Kin. 2012;44: 507–513.

[15] Lipkowski J, Ross PN. (eds.) Adsorption Molecules at Metal Electrodes. New York: VCH Publishers; 1992.

[16] Ibach H. Physics of Surfaces and Interfaces. Berlin: Springer-Verlag; 2006.

[17] Sanecki P, Skitał P, Kaczmarski K. Numerical modeling of ECE-ECE and parallel EE-EE mechanisms in cyclic voltammetry. Reduction of 1,4-benzenedisulfonyl difluoride and 1,4-naphthalenedisulfonyl difluoride. Electroanalysis 2006;18: 981–991.

[18] Sanecki P, Skitał P, Kaczmarski K. An integrated two phases approach to Zn^{2+} ions electroreduction on Hg. Electroanalysis 2006;18: 595–604.

[19] Sanecki P, Amatore Ch, Skitał P. The problem of the accuracy of electrochemical kinetic parameters determination for the ECE reaction mechanism. J. Electroanal. Chem. 2003;546: 109–121.

[20] Sanecki P, Kaczmarski K. The Voltammetric Reduction of Some Benzene-sulfonyl Fluorides, Simulation of its ECE Mechanism and Determination of the Potential Variation of Charge Transfer Coefficient by Using the Compounds with Two Reducible Groups. J. Electroanal. Chem. 1999;471: 14–25. Erratum published in J. Electroanal. Chem. 2001;497: 178-179.

[21] Sanecki P. A numerical modelling of voltammetric reduction of substituted iodobenzenes reaction series. A relationship between reductions in the consecutive-mode multistep system and a multicomponent system. Determination of the potential variation of the elementary charge transfer coefficient. Comput. Chem. 2001;25: 521–539.

[22] Sanecki P, Skitał P. The cyclic voltammetry simulation of a competition between stepwise and concerted dissociative electron transfer. The modeling of alpha apparent variability. The relationship between apparent and elementary kinetic parameters. Comput. Chem. 2002;26: 297–311.

[23] Sanecki P, Skitał P. The Application of EC, ECE and ECE-ECE Models with Potential Dependent Transfer Coefficient to Selected Electrode Processes. J. Electrochem. Soc. 2007;154: F152–F158.

[24] Sanecki P, Skitał P. The electroreduction of alkyl iodides and polyiodides The kinetic model of EC(C)E and ECE-EC(C)E mechanisms with included transfer coefficient variability. Electrochim. Acta 2007;52: 4675–4684.

[25] Skitał P, Sanecki P. The ECE Process in Cyclic Voltammetry. The Relationships Between Elementary and Apparent Kinetic Parameters Obtained by Convolution Method. Polish J. Chem. 2009;83: 1127–1138.

[26] Sanecki P, Skitał P, Kaczmarski K. The mathematical models of the stripping voltammetry metal deposition/dissolution process. Electrochim. Acta 2010;55: 1598–1604.

[27] Skitał P, Sanecki P, Kaczmarski K. The mathematical model of the stripping voltammetry hydrogen evolution/dissolution process on Pd layer. Electrochim. Acta 2010;55: 5604–5609.

[28] Speiser B. Numerical simulation of electroanalytical experiments: recent advance in methodology. In: Bard AJ, Rubinstein I. (eds.) Electroanalytical Chemistry, A Series of Advances. New York: Marcel Dekker, Inc.; 1996. vol. 19.

[29] Britz D. Digital Simulation in Electrochemistry. Berlin: Springer; 2005.

[30] Gosser JrDK. Cyclic Voltammetry: Simulation and Analysis of Reaction Mechanisms. New York: VCH Publishers, Inc.; 1993.

[31] Bard AJ, Faulkner LR. Electrochemical Methods, Fundamentals and Aplications. New York: Wiley; 2001.

[32] Bieniasz LK. Towards Computational Electrochemistry - A Kineticist's Perspective. In: Modern Aspects of Electrochemistry. Conway BE, White RE. (eds) New York: Kluwer Academic Publishers; 2002. vol35. p135-195.

[33] Bieniasz LK, Britz D. Recent Developments in Digital Simulation of Electroanalytical Experiments. Polish J. Chem. 2004;78: 1195–1219.

[34] Lasia A, Bouderbala M. Mechanism of Zn(II) reduction in DSO on mercury. J. Electroanal. Chem. 1990;288: 153–164.

[35] Niki KK, Hackerman N. The effect of normal aliphatic alcohols on electrode kinetics. J. Phys. Chem. 1969;73: 1023–1029.

[36] Niki KK, Hackerman N. Effect of n-amyl alcohol on the electrode kinetics of the V(II)/V(III) and Cr(II)/Cr(III) systems. J. Electroanal. Chem. 1971;32: 257–264.

[37] Andreu R, Sluyters-Rehbach M, Remijnse AG, Sluyters JH. The mechanism of the reduction of Zn(II) for $NaClO_4$ base electrolyte solutions at the DME. J. Electroanal. Chem. 1982;134: 101–115.

[38] Van Venrooij TGJ, Sluyters-Rehbach M, Sluyters JH. Electrode kinetics and the nature of the metal electrode. The Zn(II)/Zn electrode reaction studied at dropping gallium and mercury (micro) electrodes. J. Electroanal. Chem. 1996;419: 61–70.

[39] Hush NS, Blackledge J. Mechanism of the Zn^{II}/Zn(Hg) exchange: Part 1: the Zn^{2+}/Zn(Hg) exchange. J. Electroanal. Chem. 1963;5: 420–434.

[40] Manzini M, Lasia A. Kinetics of electroreduction of Zn^{2+} at mercury in nonaqueous solutions. Can. J. Chem. 1994;72: 1691–1698.

[41] Fawcett WR, Lasia A. Double layer effects in the kinetics of electroreduction of zinc(II) at mercury in dimethylformamide and dimethylsulfoxide solutions. J. Electroanal. Chem. 1990;279: 243–256.

[42] Barański AS, Fawcett WR. Electroreduction of Alkali Metal Cations. Part 2. Effects of Electrode Composition. J. Chem. Soc. Faraday Trans. 1. 1982;78: 1279–1290.

[43] Fawcett R, Jaworski JS. Electroreduction of Alkaline-earth Metal Cations at Mercury in Aprotic Media. J. Chem. Soc. Faraday Trans. 1. 1982;78: 1971–1981.

[44] Golędzinowski M, Kišová L. Mechanizm rozładowania jonów Cd2+ z wodnych roztworów niekompleksujących elektrolitów w obecności i nieobecności zaadsorbowanych na elektrodzie substancji organicznych. In: Galus Z. (ed.) Adsorption on Electrodes and Inhibition of Electrode Reactions. Warsaw-Lodz: The materials of 4th Symposium of Polish Chemical Society; 1980. p75 (in Polish).

[45] Ikeda O, Watanabe K, Taniguchi Y, Tamura H. Adsorption effects of highly polarizable organic compounds on electrode kinetics. Bull. Chem. Soc. Jap. 1984;67: 3363–3367.

[46] Hurlen T, Eriksrud E. Kinetics of the Zn(Hg)/Zn(II) electrode in acid chloride solution. J. Electroanal. Chem. 1973;45: 405–410.

[47] Van der Pol F, Sluyters-Rehbach M, Sluyters JH. On the elucidation of mechanisms of electrode reactions by combination of A.C. and faradaic rectification polarography. Application to the Zn^{2+} /Zn(Hg) and Cd^{2+} /Cd(Hg) reduction. J. Electroanal. Chem. 1975;58: 177–188.

[48] Pérez M, Baars A, Zevenhuizen SJM, Sluyters-Rehbach M, Sluyters JH. Establishment of an EEC mechanism for the Zn^{2+} /Zn(Hg) electrode reaction. A dropping zinc amalgam microelectrode study. J. Electroanal. Chem. 1995;397: 87–92.

[49] Tamamushi R, Ishibashi K, Tanaka N. Polarographic study on the electrode reaction of zinc ion. Z. Physik. Chem. N. F. 1962;35: 209–221.

[50] Bhugun I, Savéant J-M. Self-inhibition in catalytic processes: cyclic voltammetry. J. Electroanal. Chem. 1996;408: 5–14.

[51] Szklarczyk M. Photoelectrocatalysis. In: Murphy OJ., Srinivasan S, Conway BE. (eds.) Electrochemistry in Transition. From the 20th to 21st Century. New York and London: Plenum Press; 1992. Chapter 15. p206 - Fig. 1 and p207- Fig. 3.

[52] Guiochon G, Lin B. Modeling for Preparative Chromatography. Amsterdam: Academic Press; 1994. Chapter IV.

[53] de Levie R. Stochastics, the basis of chemical dynamics. J. Chem. Educ. 2000;77: 771–774.

[54] Souto RM, Sluyters-Rehbach M, Sluyters JH. On the catalytic effect of thiourea on the electrochemical reduction of cadmium(II) ions at the DME from aqueous 1 M KF solutions. J. Electroanal. Chem. 1986;201: 33–45.

[55] Meites L. Polarographic Techniques. second edition. New York: John Wiley & Sons, Inc.; 1965.Chapter 6.

Chapter 3

Electrochemical Basis for EZSCAN/SUDOSCAN: A Quick, Simple, and Non-Invasive Method to Evaluate Sudomotor Dysfunctions

Hanna Ayoub, Jean Henri Calvet, Virginie Lair,
Sophie Griveau, Fethi Bedioui and Michel Cassir

Additional information is available at the end of the chapter

http://dx.doi.org/10.5772/53965

1. Introduction

Globally, as of 2010, an estimated 285 million people had diabetes, with type 2 making up about 90% of the cases. Its incidence is increasing rapidly, and by 2030, this number is estimated to almost the double. Diabetes mellitus occurs throughout the world, but is more common (especially type 2) in the most developed countries. The greatest increase in prevalence is, however, expected to occur in Asia and Africa, where most patients will probably be found by 2030. The increase in incidence in developing countries follows the trend of urbanization and lifestyle changes, perhaps most importantly a "Western-style" diet [1].

All forms of diabetes increase the risk of long-term complications. These typically develop after many years (10–20), but may be the first symptom among those which have otherwise not received a diagnosis before that time. The major long-term complications relate to damage to blood vessels. Diabetes doubles the risk of cardiovascular diseases. The main "macrovascular" diseases (related to atherosclerosis of larger arteries) are ischemic heart disease (angina and myocardial infarction), stroke and peripheral vascular disease. Diabetes also causes "microvascular" complications as damage to the small blood vessels [2]. Diabetic retinopathy, which affects blood vessel formation in the retina of the eye, can lead to visual symptoms, reduced vision, and potentially blindness. Diabetic nephropathy, the impact of diabetes on the kidneys, can lead to scarring changes in the kidney tissue, loss of small or progressively larger amounts of protein in the urine, and eventually chronic kidney disease requiring dialysis. Diabetic neuropathy is the impact of diabetes on the nervous system, most commonly causing numbness, tingling and pain in the feet and also increasing the risk of skin damage due to altered sensation. Together with vascular disease in the legs, neuropathy contributes to the risk of diabetes-related foot problems (such as diabetic foot ulcers) that can be difficult to treat and occasionally require amputation. Peripheral neuropathy is the most prevalent complication of type 2 diabetes. The 2004 National Health and Nutrition Examination Survey (NHANES) on lower extremity complications revealed that close to 10% of people with diabetes have peripheral arterial disease, but close to 30% have neuropathy. The survey also showed that over 7% have an active foot ulcer, a frequent cause of hospitalization and a common pathway to amputation [3].

Peripheral neuropathy is often recognized by patients or their physicians at a time when symptoms outweigh physical signs. Sensory symptoms, paresthesias, sensory loss, and neuropathic pain are common initial complaints. Although injury to small fiber calibers and types occurs, small-diameter unmyelinated or lightly myelinated nociceptive and autonomic fibers are often prominently affected in these common neuropathies. There is an increasing interest in recognizing and treating neuropathy early in its course. Sweat glands are innervated by the sudomotor, postganglionic, unmyelinated cholinergic sympathetic C-fibers that are thin and can be damaged very early in diabetes. Sudomotor dysfunction may result in dryness of foot skin and has been associated with foot ulceration. Assessment of sudomotor dysfunction contributes to the detection of autonomic dysfunction in diabetic peripheral neuropathy and American Diabetes Association suggests that sudomotor function assessment of small fiber status should be included in the diagnostic tests for the detection of neuropathies in diabetes. The quantitative sudomotor axon reflex test (QSART) is capable of detecting distal small fiber polyneuropathy and may be considered the reference method for the detection of sudomotor dysfunction [4]. Other available techniques for assessment of sudomotor function include the thermoregulatory sweat test, silastic imprint method, the indicator plaster method. But these tests are time consuming and highly specialized tests so they are mainly used for research purposes. Skin biopsies have been developed to assess small C-Fibers. Recently, this method has been improved by assessment of small C-Fibers that innervate sweat glands. However, this method is invasive [5].

The management of diabetes alone renders considerable expenditure, however macrovascular and microvascular complications are the major cause of healthcare costs. Data from a United States study indicate that renal and cardiovascular complications seem to be the most prevalent and are associated with particularly high costs. In this study abnormal renal function and end-stage renal disease were shown to vastly increase costs of diabetes treatment, up to 771%. Furthermore, an analysis of several individual studies showed that atherosclerosis in Type 2 diabetes accounted for approximately one third of the total healthcare costs related to the disease. In a study of patients with myocardial infarction (MI), patients with diabetes had a higher per-patient total direct medical charge (inclusive of initial hospitalisation) compared to patients without diabetes.

In 2000, it was estimated that 25% of overweight adults aged 45-74 years had prediabetes, which translates into about 12 million persons in the U.S. Furthermore, in the process of identifying those with prediabetes, it was estimated that an additional 6.5 million persons with undiagnosed diabetes would have been detected. Recent controlled trials on diabetes prevention have confirmed that lifestyle changes such as diet, weight loss, and exercise as well as the drug metformin can substantially delay or prevent the progression from impaired metabolism to type 2 diabetes. Thus, several million individuals could benefit from diabetes prevention intervention [6].

There is a need to diagnose subjects with high risk for diabetes or cardiovascular diseases at an earlier stage using a simple, non-invasive, sensitive, quick and inexpensive tool. Presently, prediabetes and diabetes are diagnosed by blood glucose or HbA_{1C} levels with threshold values that were initially based on risk for retinopathy, a microvascular complication. Based on studies performed with QSART or skin biopsies, small fiber neuropathy has been shown to develop early in patients with prediabetes or cardiometabolic risk. However, these two highly specialized methods cannot be used for a large screening.

2. Principle of the method

EZSCAN/SUDOSCAN is a device (Figure 1) recently developed to provide an accurate evaluation of sweat gland function [7]. Patients place their hands and feet on electrodes, placed on skin region with a high density of sweat glands, and an incremental low direct voltage (lower than 4 V) is applied during a two minute interval. Electro-sweat conductance (ESC) is then calculated from the resulting voltage and the generated current, which is expressed in three ways: (i) current as a function of the anodic potential (so-called E), (ii) current as a function of the absolute values of the cathodic potential after applying an incremental voltage at the anode (so-called V), and (iii) current as a function of U = E+V. It is a dynamic method allowing evidence of sweat dysfunction not detectable in physiological conditions. Quantitative results are expressed as ESC, in microsiemens, μS) for the hands and feet, and a risk score is derived from the ESC values and demographic data.

3. Clinical studies for the evaluation of the performances of the device

Several studies were performed to demonstrate the robustness of the method followed by a proof of concept study and to validate the use of the EZSCAN/SUDOSCAN in detection of diabetes complications and in screening of prediabetes [8].

a. Symmetry

As the commonest form of diabetic neuropathy is symmetric, it was important to ensure that ESC measurements between right and left side were comparable. In this way, ESC in hands and feet were compared between right and left side to assess agreement between both sides using a Bland-Altman plot [9]. Coefficient of variation calculated on 1365 subjects was 3% for hands and 2% for feet, between right and left side.

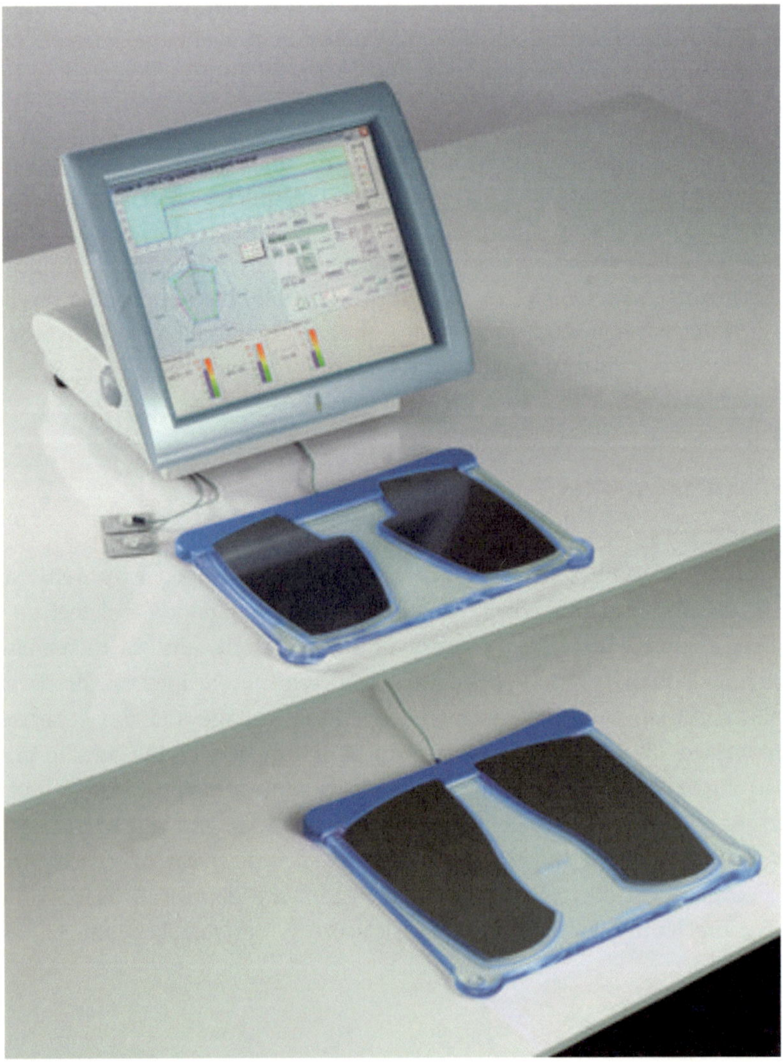

Figure 1. Photo of the device: screen and electrodes.

b. Gender effect

No significant difference was observed in ESC measured in hands and feet between female and male subjects involved in the studies or surveys performed [8].

c. Reproducibility

Measurements were assessed twice in the same day in patients with at least one cardiovascular risk and in patients with diabetes. Results were compared using a Bland-Altman plot [9]. The coefficient of variation was 7% in hands and 5% in feet in patients with cardiovascular risk and 15% in hands and 7% in feet in patients with diabetes. Coefficient of variation for glycemia, which is a gold standard for diabetes, between the two measurements was 32%.

d. Effects of glycemia

This technology has to be used in patients with prediabetes or diabetes, with potential high variations in glycemia. Thus, it was important to ensure that measurements were not influenced directly by glycemia itself. Thus, measurements were performed with a set of ten patients when their glycemia was greater than 18 mM/L and compared with measurements performed in the same patients when glycemia was below 6 mM/L. Coefficient of variation from a Bland and Altman plot with or without hyperglycemia was 10% for foot ESC.

e. Diagnosis of diabetes complications

- Screening of peripheral neuropathy

Perturbation of pain sensation is considered as one of the major initiating risk factors for diabetic foot ulcer. Sweat dysfunction leading to abnormal skin conditions including dryness and fissures could increase the risk of foot ulcers. The aim of this study was to evaluate SUDOSCAN as co-indicator of severity of diabetic polyneuropathy. 142 patients with diabetes (age 62 ± 18 years, diabetes duration 13 ± 14 years, HbA_{1c} 8.9 ± 2.5%) were measured for vibration perception threshold (VPT) using a biothesiometer and for sudomotor dysfunction by measuring ESC. Feet ESC showed a descending trend from 66 ± 17 µS to 43 ± 39 µS corresponding to an ascending trend of VPT threshold from < 15 V to > 25 V ($p = 0.001$). Correlation between VPT and ESC was -0.45 ($p < 0.0001$). Foot ESC was lower in patients with fissures while VPT was comparable. Both VPT and foot ESC were correlated with retinopathy status [10].

- Screening of cardiovascular autonomic neuropathy

Cardiovascular Autonomic Neuropathy (CAN) is a common but overlooked complication of diabetes. SUDOSCAN was compared to Heart Rate Variability (HRV) and to Ewing tests, known to be reliable methods for the investigation of CAN.

232 patients with diabetes were measured for HRV at rest and during moderate activity (stair climbing). Time and frequency domain analysis techniques, including measurement of Standard Deviation of the average beat to beat intervals (SDNN) over 5 minutes, High Frequency domain component (HF) and Low Frequency domain component (LF), were assessed during HRV testing. Heart rate variations during deep breathing and heart rate and blood pressure responses while standing, as described by Ewing according to the recommendations of the French Health Authority were also assessed. ESC was measured on the hands and feet, and a risk score was calculated. Patients were classified according to their risk score. The classifications were as follow: no sweat dysfunction, moderate sweat dysfunction and high sweat dysfunction. All results are means ± SD.

The highest correlation was observed between the risk score based on sudomotor function and the LF component during moderate activity ($r = 0.47$, $p < 0.001$). The risk score was higher in patients with a LF component value during moderate activity of < 90 ms^2 (1st quartile) when compared to LF > 405 ms^2 (3rd quartile) (46 ± 13 vs. 30 ± 13, $p < 0.001$). The risk score based on sweat function was higher in patients with 2 abnormal Ewing tests when compared to patients with all tests normal (47 ± 12 vs. 34 ± 14, $p < 0.001$). When taking two abnormal Ewing tests as reference the AUC (air under curve) of the Receiver Operating Characteristic (ROC) curve for this risk score was 0.74 with a sensitivity of 92% and a specificity of 49% for a risk score cut-off value of 35%. Regarding ROC curve analysis when choosing LF power component during moderate activity at a threshold of 90 ms^2 (1st quartile) as reference, the AUC was higher for SUDOSCAN risk score (0.77) compared to standards Ewing tests : E/I ratio (0.62), 30:15 ratio (0.76) and blood pressure change to standing (0.65). Using a cut-off of 35%, for SUDOSCAN risk score sensitivity and specificity were respectively 88% and 54% [11].

- Screening of diabetic nephropathy

Given the inter-relationships between dysglycemia, vasculopathy and neuropathy, it was hypothesized that SUDOSCAN may detect diabetic kidney disease (DKD).

In a case-control cohort consisting of 50 Chinese type 2 diabetic patients without DKD (ACR < 2.5mg/mM in men or ACR < 3.5mg/mM in women and eGFR > 90 ml/min/1.73 m^2) and 50 with DKD (ACR ≥ 25 mg/mM and eGFR < 60ml/min/1.73 m^2), we used spline analysis to determine the threshold value of SUDOSCAN score to predict DKD and its sensitivity and specificity.

SUDOSCAN scores were highly correlated with log values of eGFR ($r = 0.67$, $p < 0.0001$, see Figure 2) and ACR ($r = -0.66$, $p < 0.0001$). Using a cutoff value of 55 on the risk score scale, the score had 94% sensitivity and 78% specificity to predict DKD with a likelihood ratio of 4.2, positive predictive value of 81% and negative predictive value of 93%. In patients without DKD, those with low SUDOSCAN score (n = 10) had longer disease duration [median (IQR): 13 (9-17) vs. 8 (4-16) years, p=0.017] and were more likely to have retinopathy (36.7% vs. 5.1%, p=0.02), lower eGFR [98 (95.00-103) vs. 106 (98.5-115), p = 0.036] and more treated with RAS blockers (81.8% vs. 25.6%, p = 0.002) than those with normal score. On multivariable analysis, SUDOSCAN score remained an independent predictor for DKD (1= yes, no = 0) ($\beta = -0.72$, $p = 0.02$) along with smoking ($\beta = -2.37$, $p = 0.02$), retinopathy ($\beta = 3.019$, $p = 0.01$), triglyceride ($\beta = 2.56$, $p = 0.013$) and blood hemoglobin ($\beta = -0.613$, $p = 0.04$) [12].

f. Proof of concept study

As sweat chloride movements in sweat ducts are likely to be impaired in cystic fibrosis (CF), SUDOSCAN results were compared in CF patients and control subjects. ESC, measured when a very low voltage is applied, and dESC, difference between ESC at very low voltage and ESC at low voltage, were assessed in 41 adult patients with classical CF and 20 healthy control subjects.

ESC measurements on hands and feet were significantly higher in CF patients as compared to control subjects. dESC was significantly lower in CF patients and more discriminative (9 ± 18 μS vs. 49 ± 31 μS, $p < 0.0001$ on hands and 34 ± 24 μS vs. 93 ± 24 μS, $p < 0.0001$ on feet) (see Figure 3 for an individual comparison). dESC measurement provided a diagnostic specificity of 1 and a sensitivity of 0.93. Correlation between feet ESC and sweat chloride concentration as measured by sweat test was -0.70 ($p < 0.0001$). Precision for the measurements was 6% for hands ESC, 4% for feet ESC, 9% for hands dESC and 3% for feet dESC [13].

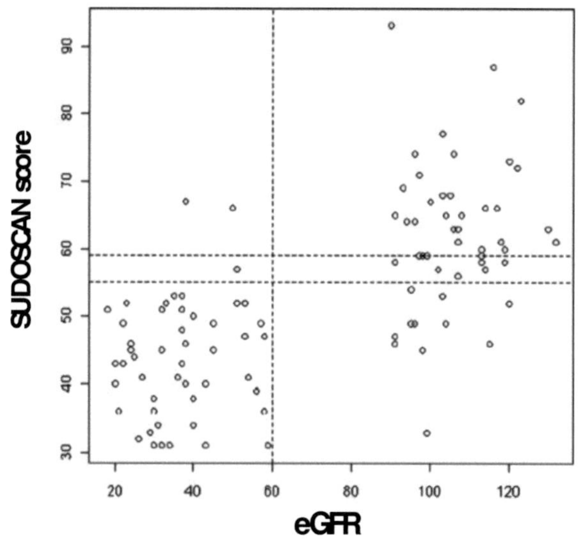

Figure 2. A scatter plot and grid analysis showing the correlations between estimated glomerular filtration rate (eGFR) and EZSCAN scores [adapted from 12].

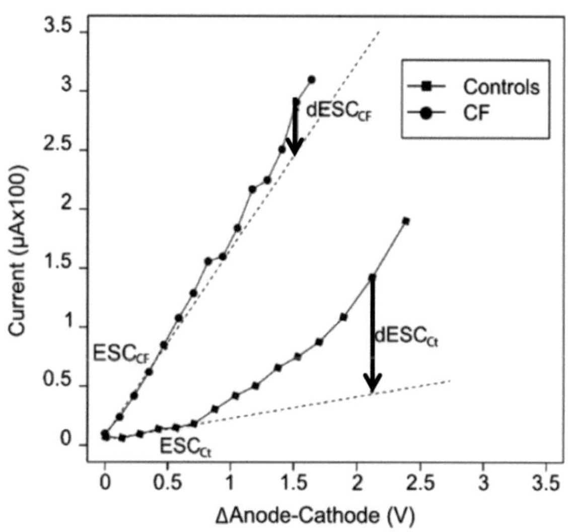

Figure 3. Individual current-voltage current curves for a control subject and a patient with cystic fibrosis showing dESC, the difference between ESC at low voltage and at high voltage dESC is shown by the up down arrow.

g. Identification of subjects at high risk of diabetes

A longitudinal study was performed in subjects with an initial normal glucose tolerance (NGT) to assess the ability of EZSCAN to predict future abnormalities in glucose tolerance.

South Asian (Indian) subjects (n = 69, 48% male, mean age 42 ± 9 years, mean BMI 28 ± 5 kg/m^2) diagnosed as NGT with a previous oral glucose tolerance test (OGTT, T0) underwent a frequently sampled OGTT (FSOGTT), 8 months later (T8) with calculation of the area under the curve (AUC) for glucose and insulin. At both times EZSCAN tests were done. Using $AUC_{glucose}$ and $AUC_{insulin}$ measured by the FSOGTT, subjects were categorised as normal, high $AUC_{insulin}$ or isolated high $AUC_{glucose}$. Odds ratio (OR) for having high $AUC_{insulin}$ or isolated high $AUC_{glucose}$ vs. normal was computed by logistic regression analysis using EZSCAN risk classification at T0 as independent variable (< 50% = normal, no risk, 50-65% = intermediate risk and > 65% = high risk).

At T8, 11 and 5 subjects developed impaired glucose tolerance and diabetes respectively. OR of having high $AUC_{insulin}$ or isolated high $AUC_{glucose}$ in the different risk groups was 6.19 (CI 95% 1.50– 25.48, p = 0.0116) for high risk vs. no risk and 3.0 (CI 95% 0.98–9.19, p = 0.0545) for intermediate risk vs. no risk. Sensitivity of EZSCAN for early detection of these abnormalities in glucose tolerance was 77% while it was 14% for fasting plasma glucose and 66% for HbA_{1C} [14].

h. Conclusion on clinical studies

All these clinical studies that have been published in international peer reviewed journals evidenced that sweating status as assessed by EZSCAN/SUDOSCAN:

- is a robustness method with good reproducibility
- is a sensitive method when compared with the conventional methods may be very useful to identify and manage subjects at risk for developing glucose intolerance
- may be a quantitative indicator on the severity of polyneuropathy that may be useful for the early prevention of foot skin lesions
- may be used for the early screening of cardiovascular autonomic neuropathy in daily clinical practice before more sophisticated, specific, and time-consuming tests
- may be used to detect high risk subjects for diabetic kidney diseases

This quick and simple method is well accepted by the subject, does not require specific preparation and does not need high training allowing its performance by non specialized teams.

4. *In-vitro* study

Experiments were performed *in-vitro* to improve the method and to understand: i) the role of the components of the sweat in electrochemical reaction with nickel; ii) the onsets of currents observed in clinical studies and the influencing factors; iii) the electrochemical kinetics of reactions; iv) the effect of ageing of electrodes; v) the consequences on electrochemical reaction of the use of stainless-steel electrodes generally recommended for medical use.

4.1. Electrochemical characterization of nickel electrodes in phosphate and carbonate electrolytes

Although the electrochemical properties of nickel have been widely examined, through the analysis of its corrosion in aqueous acid or alkaline solution, very few studies have been dedicated to the specific assessment of its behavior in physiological solutions. In a preliminary study, we thoroughly explored the electrochemical behavior of nickel electrode (i) in a three-electrode set-up combining a nickel counter electrode and a nickel pseudo-reference electrode in order to mimic the whole Ni electrode configuration of the SUDOSCAN™ device; (ii) and in synthetic buffered phosphate and carbonate solutions (PBS and CBS) in which the pH and the concentrations of chloride, lactate and urea were varied to mimic the behavior of the electrodes in contact with sweat.

This approach provides insight into the origin of the onset of responses measured upon the application of low voltage potential with variable amplitudes to Ni electrodes. This study also constitutes a sound approach aimed at understanding the chemical key parameters controlling the electrochemical currents.

- **Anodically:** for low voltage amplitude, the electrochemical reactions measured at the electrodes are mainly those related to the oxidation of Ni leading to the formation of an oxy-hydroxide film [15]. The following reactions have been proposed for Ni oxidation in presence of Cl^- [16,17]:

$$Ni + H_2O = Ni(H_2O)_{ad} = Ni(OH)_{ad} + H_{aq}^+ + e^-$$

$$Ni(H_2O)_{ad} + Cl^- = Ni(ClOH)^-_{ad} + H^+ + e^-$$

$$Ni(OH)_{ad} + H^+ = Ni^{2+}{}_{aq} + (H_2O)_{ad} + e^- = Ni(OH)_2$$

At high potentials, the breakdown of the oxy-hydroxide film becomes the main anodic reaction [15]. This is mainly due to the competitive adsorption between Cl^-, oxygen containing species, carbonate anions and the formation of soluble species as Ni-Cl- and/or NiO(H)-Cl- leading to the breakdown of the oxy-hydroxide film. This could also be due to the direct penetration of chloride ions through the oxy-hydroxide film. Therefore, the increase in the concentration ratios $[Cl^-]/[OH^-]$ acts in favor of the adsorption of Cl^- and, thus, the weakness of the passive layer occurs, leading to its breakdown at low potential values.

- **Cathodically:** for low voltage amplitude, the electrochemical reactions are mainly related to the reduction of the oxy-hydroxide film. The following reactions have been proposed:

$$NiO + 2H^+ + 2e^- \rightarrow Ni + H_2O$$

And/or

$$Ni(OH)_2 + 2H^+ + 2e^- \rightarrow Ni + 2H_2O$$

At high voltage potential, the reduction of the oxy-hydroxide film and the electrolytic solution govern the cathode reactions [15].

In the particular case of CBS (36 mM, pH 6.4) containing different concentrations of chloride ions, within the expected range of Cl^- concentration in sweat, the positive potential going direction of the cyclic voltammograms (Figure 4) show that in all cases, an anodic plateau appears at ca. 0.3 V, indicating the formation of a passive film composed probably by $Ni(OH)_2$ and/or NiO.

In this examined range of potentials, the voltammograms show also a large anodic current at high potentials due to the localized dissolution of the nickel following Cl^- attack. The increase in the concentration of Cl^- initiates earlier the localized dissolution of Ni. Indeed, the breakdown potential "E_b" shifts towards lower anodic potentials when increasing the concentration of Cl^-.

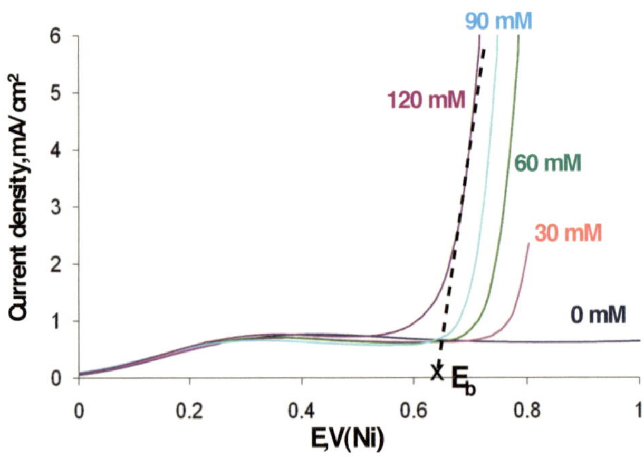

Figure 4. Cyclic voltammograms of Ni electrode in aerated CBS (36 mM; pH 6.4) in presence of different concentration of NaCl. Scan rate: 100 mV/s [adapted from 15].

Moreover, the variation of buffer, urea and lactate concentrations does not have a significant effect on the electrochemical anodic behavior of Ni and notably on the breakdown potential "E_b". Therefore, the anodic currents are likely to be controlled by the variation of Cl^- concentration [15,18].

The obtained results (data not show) indicate that the cathodic currents are less affected by the variation of the electrolyte concentrations and they are likely to be controlled by the variation of pH value.

4.2. Comparison of *in-vitro* results and clinical observations

In order to establish a parallel between the *in-vitro* observations and those obtained during the clinical tests to further understand the origin of the onsets of currents and their evolution with chloride ion concentrations, linear anodic voltammograms were performed. The induced potentials on the counter electrode (playing the role of cathode in this case) were measured simultaneously. The results are then expressed, as in the medical technology (Figure 5), as follows: (1) I vs. E, (2) I vs. V and (3) I vs. U (= E+V) where I is the current.

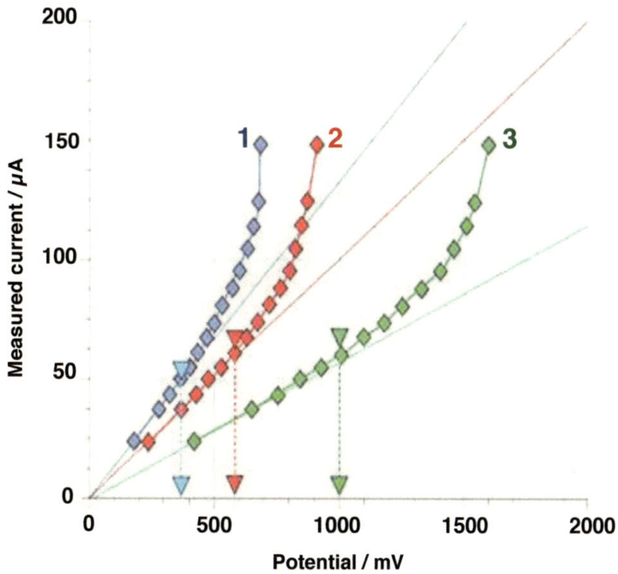

Figure 5. Example of the electrochemical results obtained by the SUDOSCAN technology (1: I vs. E; 2: I vs. V; and 3: I vs. U = E+V).

As shown above, the variation of chloride concentration is the main sweat parameter affecting the anodic electrochemical behavior of Ni. This led us to study the influence of the concentrations of Cl^- on the measured variation of current-voltage outputs. Linear anodic voltammograms were performed in (I) -0.3 V to 0.9 V vs. SCE potential range and the induced potentials on the counter electrode (playing the role of cathode in this case) were measured simultaneously. The results are then expressed, as in SUDOSCAN™ technology.

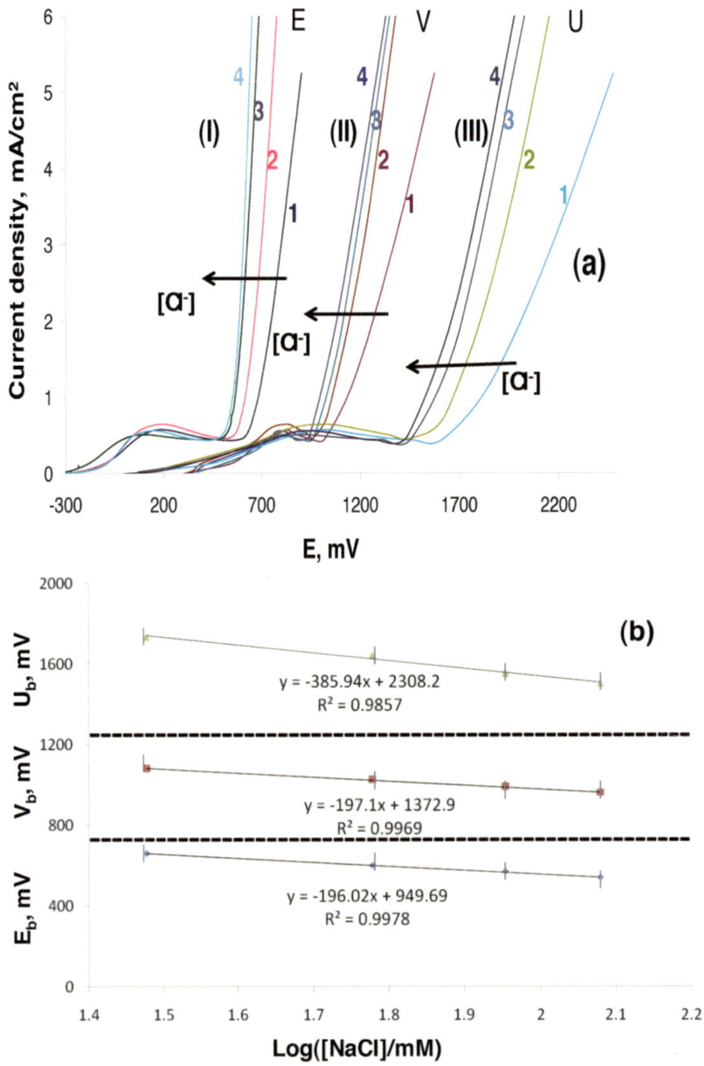

Figure 6. a) (I) j vs. E, (II) j vs. V, and (III) j vs. U (= E+V), where j is the current density, after sweeping the potential between -0.3 V → 0.8 V vs. SCE in CBS (36 mM, pH 7) and in presence of NaCl (curves 1: 30 mM, curves 2: 60 mM, curves 3: 90 mM, and curves 4: 120 mM). (b) Evolution of E_b, V_b and U_b as a function of Log[Cl^-] (data from Figure 6a) [adapted from 19].

In the particular case of pH 7, Figure 6a depicts the results obtained in CBS (36 mM, pH 7), containing different concentrations of NaCl (30, 60, 90, and 120 mM). It clearly appears that the concentration of Cl^- affects the variation of the current as a function of E, V and U. In fact, the increase in the concentration of Cl^- shifts linearly the breakdown potential, E_b, towards more cathodic values, as it can be seen in Figure 6b. According to the same way than E_b, V_b and U_b can be also evaluated as the potentials associated to the point of deviation of j-V and j-U curves. Figure 6b shows that the increase in the concentration of Cl^- shifts linearly

E_b and V_b towards lower V values and the slope value of V_b vs. Log[Cl^-] is close to that of E_b vs. Log[Cl^-]. Consequently, the increase in Cl^- concentration shifts linearly U_b towards lower U values and the slope of U_b vs. Log[Cl^-] is about twice the value found for E_b or V_b vs. Log[Cl^-]. It should be noted that the results obtained at pH 6 provided the same features as in the present case. This implies that the evolution of the current-voltage curves reported in Figures 6a and 6b are not very sensitive to the variation of pH values.

Furthermore, curves obtained for generated current as a function of E, V, U during clinical tests are quite comparable (Figure 5 and 6a). These results led us to conclude that the determination of the current curve as a function of the potential (anode, cathode or their difference) provides a very efficient way to detect the deviation in the ion balance and notably the deviation in Cl^- concentration at the level of the electrodes.

4.3. Electrochemical kinetics of anodic nickel dissolution

As shown above, the variation of chloride ion concentration plays a key role in predicting sudomotor dysfunction by controlling the generated current at high anodic potentials, notably the currents related to the localized anodic dissolution of nickel. This led us to study the kinetics of the different electrochemical reactions related to the localized dissolution of nickel in carbonate buffer solutions (CBS) at different physiological pH values and in presence of different concentrations of chloride ions, within the expected range of concentrations in sweat at rest. This aims at studying the mechanisms of the electrochemical reactions and completing a theoretical model on the basis of the electrical signals registered during the clinical tests. Our results show that in a pH range between 5-7, the rate determining step appears to be the transfer of a first one-electron, as suggested by Tafel slopes close to 0.120 V/decade. However, the reaction order in chloride ions changes from around 2, for pH 7, to around 1, for pH values between 6 and 5 without a change in the rate-determining step [20].

At pH 7, the following mechanism has been proposed for the reactions taking place before and during the rate determining step (rds):

1. $Ni + 2Cl^- \leftrightarrows [NiCl_2]^{2-}{}_{ads}$
2. $[NiCl_2]^{2-}{}_{ads} \leftrightarrows [NiCl_2]^- + e^-$ rds

And at pH 5-6, the following mechanism has been proposed:

1. $Ni + Cl^-$ (aq) $\leftrightarrows Ni(Cl^-)_{ads}$
2. $Ni(Cl^-)_{ads} \leftrightarrows Ni(I)Cl + e^-$ rds

4.4. Effect of electrode ageing on measurements

The nickel electrodes play alternately the role of anode and cathode, which do not undergo any specific pretreatment before each measurement. Thus, the analysis of the temporal evolution of the physico-chemical properties of nickel is of prime importance to ensure the good performance of the medical device. The objectives of the present work are to study both the electrochemical behavior and the surface chemical composition of nickel electrodes after ageing under repeated cyclic voltammograms in different potential windows. Surface chemical characterizations by XPS (X-ray photoelectron spectroscopy) and ToF-SIMS (Time of Flight-Secondary Ion Mass Spectrometry) were performed on nickel electrodes after ageing with repeated cyclic voltammograms in carbonate buffer solutions containing the main components of sweat, at different potential ranges, in: (i) a restricted anodic potential range; (ii) a negative extended potential range.

The electrochemical behavior of Ni electrodes was first studied after ageing with repeated cyclic voltammograms in a potential range of -0.3 V to 0.5 V and return to -0.3 V (anodic region only). Then, in order to assess the influence of alternating the polarity of the electrodes during the clinical tests on the Ni ageing behavior (each electrode playing alternately the role of anode or cathode), successive cyclic voltammograms were performed in a potential range of -0.3 V → 0.5 V → -1 V (cathodic region). All the experiments were conducted in aerated CBS solutions, with a concentration of 120 mM of NaCl.

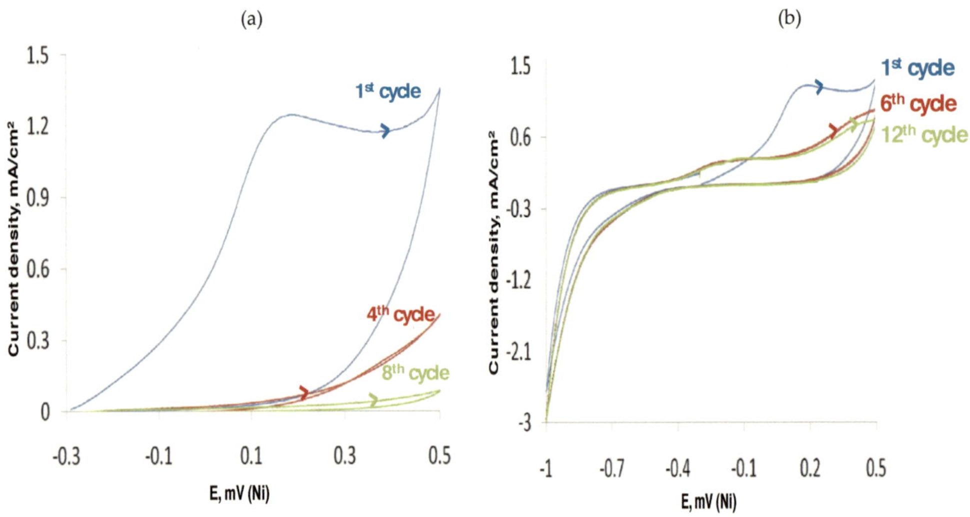

Figure 7. Successive cyclic voltammograms of Ni electrode in CBS (pH 6.4) in presence of 120 mM NaCl. Scan rate: 100 mV/s. (a) potential range: -0.3 V → 0.5 V → -0.3 V and (b) potential range: -0.3 V → 0.5 V → -1.0 V [adapted from 21].

Figure 7a shows cyclic voltammograms after 1, 4 and 8 cycles in aerated CBS (36 mM) in presence of Cl^- (120 mM) in the potential range [-0.3 V → 0.5 V → -0.3 V]. It clearly appears that the first potential sweep strongly affects the subsequent cyclic voltammograms. The anodic plateau of the first cycle, attributed to Ni oxidation, is no longer observed in the following cycles (see, for example, the voltammograms of the fourth cycle and eighth cycles in Figure 7a). These changes in the voltammograms are mainly attributed to the formation of an oxide layer firmly attached to the metal and forming a compact barrier between the metal and the solution with a very low electronic conductivity during the first cycle [22].

Figure 7b shows the obtained cyclic voltammograms after 1, 6 and 12 cycles when the potential on the return cycle is extended to -1 V (potential range [-0.3 V → 0.5 V → -1 V → -0.3 V]). Contrarily to the results presented in Figure 7a, there are two new observations: (i) the Ni oxidation process is still observed during the subsequent cycles around -0.4 V and 0.4 V, (ii) the high intensities of the anodic current remain with the subsequent cycles. This is mainly due to a partial reduction or re-activation [22,23] of the compact oxide film when the potential scan is extended down to -1 V. The re-activation of the oxide film, in the cathodic reduction step, is probably due to a surface modification or post-electrochemical re-organization of the initial deposited species [24] leading to an increase in the electronic conductivity of the surface of the Ni electrode.

It should be noted here that additional experiments, performed in the absence of Cl^-, displayed the same features as in the present case (data not shown). This implies that the voltammograms reported in Figure 7 are poorly sensitive towards the presence of Cl^-, as the potential range is well below the pitting corrosion one.

XPS characterizations were performed on the series of samples shown in Figures 7a-b. The Ni $2p_{3/2}$ core levels were systematically decomposed into the spectroscopic contributions characteristic of metallic nickel (main peak located at a BE of 852.8 ± 0.2 eV, nickel plasmon at 856.3 ± 0.2 eV and satellite at 858.8 ± 0.2 eV [25-27]), nickel oxide NiO (main peak located at 854.7 ± 0.2 eV and two satellites located at 856.4 ± 0.2 eV and 861.7 ± 0.2 eV [25-27]) and $Ni(OH)_2$ (main peak at 856.7 ± 0.2 eV and satellite at 862.6 ± 0.2 eV [26,27]). Figure 8 displays these core levels and their decomposition into individual contributions of Ni, NiO, and $Ni(OH)_2$.

Based on previously published data, a simple layer model of the passive film can be suggested. It is composed of an homogeneous continuous outermost layer of $Ni(OH)_2$ and an homogeneous continuous inner NiO oxide layer, in contact with the metal [28,29].

It was possible to calculate the equivalent thicknesses of the $Ni(OH)_2$ and NiO layers, as well as the total oxidized surface layer (as the arithmetic sum of the two former ones), from the peak intensities of the fitted Ni $2p_{3/2}$ core levels, taking into account such two-layer model for the description of the Ni oxide film (see Equations 1 and 2) in all experiments. The values are reported in Table 1.

$$d_{NiO} = \lambda_{Ni}^{NiO}.\sin\beta.\ln\left[1+\frac{D_{Ni}^{Ni}.\lambda_{Ni}^{Ni}}{D_{Ni}^{NiO}.\lambda_{Ni}^{NiO}}\cdot\frac{I_{Ni}^{NiO}}{I_{Ni}^{Ni}}\right] \tag{1}$$

$$d_{Ni(OH)_2} = \lambda_{Ni}^{Ni(OH)_2}.\sin\beta.\ln\left[1+\frac{D_{Ni}^{Ni}.\lambda_{Ni}^{Ni}}{D_{Ni}^{Ni(OH)_2}.\lambda_{Ni}^{Ni(OH)_2}}\cdot\frac{I_{Ni}^{Ni(OH)_2}}{I_{Ni}^{Ni}}\cdot\exp\left(\frac{-d_{NiO}}{\lambda_{Ni}^{NiO}.\sin\beta}\right)\right] \tag{2}$$

Where d is the layer thickness, β is the take-off angle of the photoelectrons with respect to the sample surface, λ_M^N is the inelastic mean free path of the photoelectrons coming from M in the matrix N, I_{Ni}^{Ni} is the nickel intensity for NiO in the bulk metal, $I_{Ni}^{Ni(OH)_2}$ is the nickel intensity for $Ni(OH)_2$, D_M^N is the density of M in the matrix N, the inelastic mean free paths used in this work are the following: 1.41 nm for λ_{Ni}^{Ni} [26], 1.43 nm for λ_{Ni}^{NiO} [26] and 1.19 nm for $\lambda_{Ni}^{Ni(OH)_2}$ [26].

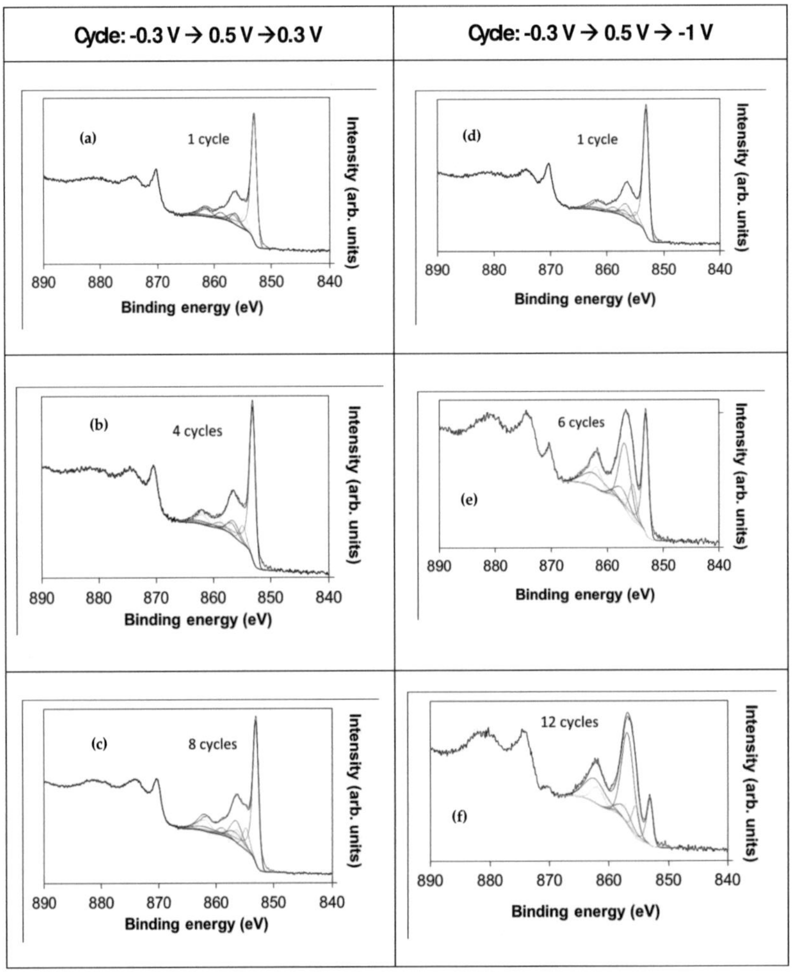

Figure 8. XPS Ni $2p_{3/2}$ core level peak decompositions of a Ni electrode immersed in carbonate buffer saline solution (pH 6.4), in presence of 120 mM NaCl, for different numbers of cycles in different potential range [adapted from 21].

In a restricted anodic potential range, XPS results indicate that the surface was passivated by a 1 nm-thick duplex layer composed of nickel hydroxide (outermost layers) and nickel oxide (inner layers). In a negative extended potential range, though the electrochemical behavior of electrodes did not change, the inner nickel oxide layer was thickening, indicating a surface degradation of the nickel electrode in these conditions.

Sample treatment	NiO equivalent thickness (nm)	$Ni(OH)_2$ equivalent thickness (nm)	Passive layer thickness (nm)
-0.3 V→ 0.5 V→ -0.3V	with 120 mM NaCl		
1 cycle	0.7 ± 0.1	0.5 ± 0.1	1.2 ± 0.2
4 cycles	0.7 ± 0.1	0.7 ± 0.1	1.4 ± 0.2
8 cycles	0.9 ± 0.1	0.7 ± 0.1	1.6 ± 0.2
-0.3V→ 0.5 V→ -1.0 V	with 120 mM NaCl		
1 cycle	0.5 ± 0.1	0.8 ± 0.1	1.3 ± 0.2
6 cycles	1.4 ± 0.1	1.3 ± 0.1	2.7 ± 0.2
12 cycles	2.6 ± 0.1	1.4 ± 0.1	4.0 ± 0.2
-0.3V→ 0.5 V→ -1.0 V	without 120 mM NaCl		
1 cycle	0.7 ± 0.1	1.0 ± 0.1	1.7 ± 0.2
6 cycles	1.8 ± 0.1	1.1 ± 0.1	2.9 ± 0.2
12 cycles	2.3 ± 0.1	1.8 ± 0.1	4.1 ± 0.2

Table 1. NiO and $Ni(OH)_2$ layer thicknesses estimated from the XPS Ni $2p_{3/2}$ core level peak decompositions (two-layer model) [adapted from 21].

These systematic observations, in different potential ranges, show that alternating the polarity of the electrodes ensures the reproducibility of measurements for a large number of clinical tests and explain why, during routine use of the medical device, the metal/sweat interaction may reduce the lifetime of the anodic and/or cathodic electrodes.

4.5. Stainless-steel electrodes behavior: Comparison with nickel

Although the contact duration of nickel with skin is only about 2 minutes, the risk of allergic reactions cannot be discarded. In order to improve the device, a new electrode material, stainless steel 304L (SS 304L), with lower Ni content, was tested in carbonate buffer solutions in the presence of various concentrations of chloride, lactate and urea to mimic the chemical composition of the sweat.

Stainless steel 304L (SS 304L), already used in surgical instruments for example, was selected as a potential substitute material. Thus, the electrochemical behavior of SS 304L was analyzed and, more particularly, its sensitivity to the variation of different parameters in sweat. This work is aimed at understanding the adequacy of stainless steel 304L to the clinical testing application. The electrochemical measurements were performed in a three-electrode set up combining a stainless steel 304 L counter electrode in order to mimic the 2 active electrodes configuration (the same material is used for the anode and the cathode) of the SUDOSCAN™ device.

As for nickel, the influence of the variation of electrolyte concentrations on the electrochemical behavior of SS 304L, mainly appears by a deviation of the generated currents at high anodic voltage potential and notably the deviation of the breakdown potential, E_b, towards less or higher anodic potentials.

The obtained results show that SS 304L is more sensitive than Ni to the variation of Cl^- concentration. Figure 9 shows the cyclic voltammograms recorded on SS 304L in carbonate buffer solutions (36 mM, pH 7) without Cl^- and with increasing concentrations of Cl^- within the expected range of concentrations in sweat at rest. In all cases, the concentration of Cl^- is high enough to cause the destruction of the passive film. Furthermore, E_b decreases with increasing chloride concentration. A high shift of about 0.42 V was observed by varying chloride concentration from 30 to 120 mM at pH 7.

As for nickel, a linear variation between E_b and Log([Cl^-]) was also observed. However, SS 304L is more sensitive than Ni to the variation of Cl^- concentration. The slope value of the linear variation of E_b as a function of Log([Cl^-]) was ≈ 0.2 for Ni and ≈ 0.67 for SS 304L (Figure 10).

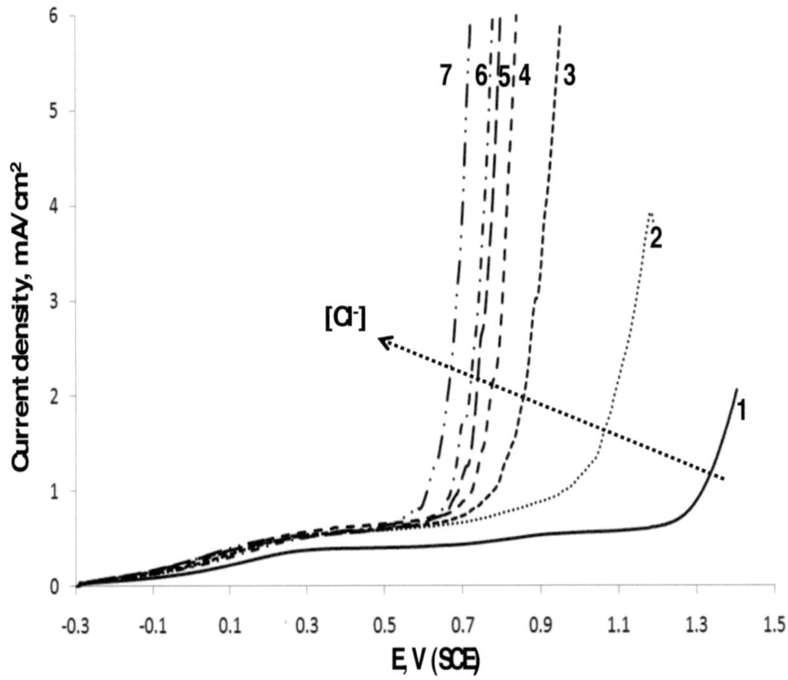

Figure 9. Cyclic voltammograms of SS 304L electrode in aerated CBS (36 mM, pH 7) in presence of NaCl (curve 1: 0 mM; curve 2: 30 mM; curve 3: 45 mM; curve 4: 60 mM; curve 5: 75 mM; curve 6: 90 mM and curve 7: 120 mM). Only the forward scans are shown. Scan rate: 100 mV/s [adapted from 30].

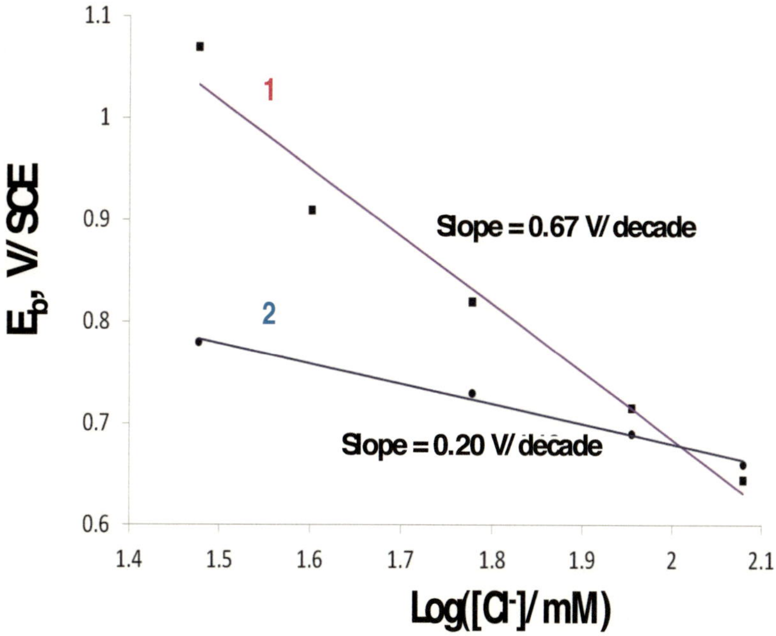

Figure 10. E_b vs. Log [Cl^-]. Data obtained from cyclic voltammograms on (1 : SS 304L electrode; 2 : nickel electrode) in carbonate buffer solutions (36 mM, pH 7) in presence of different concentrations of Cl^- (between 30 and 120 mM).

The effect of adding increasing amounts of Cl^- on E_b of SS 304L in CBS (36 mM) at pH 5, 5.5 and 6 was also examined. Table 2 presents the E_b values deduced from cyclic voltammograms recorded (data not shown) at different pH and chloride ions concentration. It clearly appears that, at pH 5 and 5.5, the localized dissolution occurs only when Cl^- concentration exceeds 40 mM, whilst at pH 6, Cl^- concentration should exceed 30 mM.

[NaCl](mM)	E_b, V_{SCE} (pH 5)	E_b, V_{SCE} (pH 5.5)	E_b, V_{SCE} (pH 6)	E_b, V_{SCE} (pH 7)
0	1.38	1.38	1.34	1.30
30	1.38	1.38	1.31	1.06
40	1.38	1.38	1.15	0.90
60	1.07	1.03	0.92	0.805
90	0.88	0.83	0.79	0.70
120	0.77	0.76	0.73	0.64

Table 2. E_b values deduced from cyclic voltammograms recorded on SS 304L in CBS of different pH and containing various chloride ions concentration [30].

The effect of pH, buffer concentration, urea concentration and lactate concentration was also studied. The obtained results show that variation of pH, buffer concentration and lactate concentration also affect, but to a less extent than chloride, the electrochemical behavior of SS 304L by displacing E_b towards lower or higher anodic potentials [30]. As the variation range of these parameters in sweat is low compared to that of Cl^-, and as the breakdown potential (E_b) is highly shifted by varying Cl^- concentration, the currents obtained during the clinical tests are likely to be controlled by the variation of Cl^- concentration. These results tend to prove that SS 304L is suitable for use in the SUDOSCAN™ application due to its high capacity to detect the deviation in the ionic balance and notably the deviation in Cl^- concentration.

5. Conclusion

The study of the electrochemical behavior of nickel was carried out in sweat-mimic solutions and using a set-up similar to that of the medical device. This study allowed us to define the origin of the onset of responses measured upon the application of low voltage potential with variable amplitudes to Ni electrodes. This study also clearly indicates that the variation of chloride concentration is the main sweat parameter controlling the electrochemical currents.

The comparisons between *in-vitro* study and clinical observations clearly indicate that the electrochemical *in-vitro* measurements on the behavior of nickel electrodes are close enough to those obtained through the clinical tests and prove that the determination of the current curve as a function of the potential (anode, cathode or their difference) provides a very efficient way to detect the deviation in Cl^- concentration, at the level of the electrodes.

The influence of chloride concentrations on the kinetics of the electrochemical reactions was also studied. The proposed mechanisms and the obtained kinetic parameters were then used by Impeto Medical to complete a theoretical model.

An evaluation of the ageing of electrodes on their performance was conducted by realizing surface analyzes, such as XPS and SIMS spectroscopies. Our results have highlighted the importance of alternating the polarity of electrodes to ensure their sensitivity and the reproducibility of measurements. However, after frequent uses, the metal/sweat interaction can lead to a slight deterioration of the electrodes surface.

Finally, in order to reduce the allergic risk, the electrochemical studies were extended to the stainless steel 304L as a replacement material of nickel. The electrochemical study shows that stainless steel 304L is very sensitive to the deviation of sweat ionic balance and notably to the variation of chloride concentrations at the level of electrodes. This makes stainless steel 304L a very promising material for the medical device application.

Results obtained from *in-vitro* studies have been used to improve the development of EZSCAN/SUDOSCAN.

http://dx.doi.org/10.5772/53965

In perspective, it could be interesting to compare the electrochemical behavior and surface modification between electrodes aged in-vivo and electrodes aged *in-vitro* and to study the electrochemical behavior of different compositions of stainless steel. It could be also interesting to study the electrochemical kinetics of anodic reactions taking place at the surface of stainless steel electrode in physiological solutions.

Author details

Hanna Ayoub[1,2,3], Jean Henri Calvet[3], Virginie Lair[1*], Sophie Griveau[2], Fethi Bedioui[2] and Michel Cassir[1]

1 LECIME CNRS UMR 7575, Chimie ParisTech, Paris, France

2 UPCGI CNRS 8151/INSERM U 1022, Université Paris Descartes, Chimie ParisTech, Paris, France

3 IMPETO Medical, Paris, France

References

[1] IDF, Diabetes Atlas Fifth Edition. Diabetes Atlas ed. IDF. Brussels: International Diabetes Federation 5 (2011)

[2] American Diabetes Association. Diagnosis and classification of diabetes mellitus. Diabetes Care 33 (2010) 62

[3] Tesfaye S, Boulton AJ, Dyck PJ, Freeman R, Horowitz M, Kemper P et al. Diabetes Care 33 (2010) 2285

[4] Low VA, Sandroni P, Fealey RD, Low PA. Muscle Nerve 34 (2006) 57

[5] Lauria G, Cornblath DR, Johansson O, McArthur JC, Mellgren SI, Nolano M, et al. Eur J Neurol 12 (2005) 1

[6] Knowler WC, Barrett-Connor E, Fowler SE, et al. Diabetes Prevention Program Research Group. Reduction in the incidence of type 2 diabetes with lifestyle intervention or metformin N Engl J Med 346 (2002) 393.

[7] Brunswick P, Bocquet N, Patent number: France 0753461 and PCT EP2008/052211.

[8] Schwarz P, Brunswick P, Calvet JH. Journal of Diabetes & Vascular diseases 11(2011) 204.

[9] Bland JM, Altman DG. Lancet 327 (1986) 307.

[10] Gin H, Baudouin R, Raffaitin C, Rigalleau V, Gonzalez C. Diabetes & Metabolism 11(2011) 527.

[11] Calvet JH, Dupin J, Deslypere JP. Journal of Diabetes & Metabolism. Accepted for publication.

[12] Ozaki R, Cheung KKT, E. Wu, A. Kong, X. Yang, E. Lau, P. Brunswick, JH. Calvet, JP.Deslypere, J.C.N. Chan. Diabetes technology & therapeutics 13 (2011) 937.

[13] Hubert D, Brunswick P, Calvet JH, Dusser D, Fajac I. Journal of Cystic Fibrosis 10 (2011)15.

[14] Ramachandran A, Moses A, Snehalatha C, Shetty S, Thirupurasundari CJ, Seeli AC. Journal of Diabetes & Metabolism 2 (2011)1.

[15] Ayoub H, Griveau S, Lair L, Brunswick P, Cassir M, Bedioui F. Electroanalysis 22 (2010) 2483.

[16] Milosev I, Kosec T, Electrochimica Acta 52 (2007) 6799.

[17] Real SG, Barbosa MR, Vilche JR, Arvia AJ, Journal of Electrochemical Society 137 (1990) 1696.

[18] Khalfallah K, Ayoub H, Calvet JH, Neveu X, Brunswick P, Griveau S, Lair V, Cassir M, Bedioui F. IEEE Sensors Journal 12 (2012) 456.

[19] Ayoub H, Lair V, Griveau S, Brunswick P, Bedioui F, Cassir M. Sensors Letters Journal 9, (2011) 2147.

[20] Ayoub H, Lair V, Griveau S, Brunswick P, Zagal J H, Bedioui F, Cassir M. Electroanalysis, 24 (2012) 386.

[21] Ayoub H, Lair V, Griveau S, Galtayries A, Brunswick P, Bedioui F, Cassir M. Applied Surface Science 258, (2012) 2724.

[22] Hoar TP. Corrosion Science 7 (1967) 341.

[23] Burke LD, Whelan DP, Electroanal J. Chem. 109 (1980) 385.

[24] Burke LD, Twomay TAM. J. Electroanal. Chem 162 (1984)101.

[25] Laksono E, Galtayries A, Argile C, Marcus P. Surface Science 530 (2003) 37

[26] Payne BP, Grosvenor A P, Biesinger M C, Kobe B A, McIntyre N S. Surf. Interface Anal. 39 (2007) 582.

[27] Biesinger M C, Payne B P, Lau LWM, Gerson A, Smart RStC. Surf. Interface Anal. 41 (2009) 324.

[28] Marcus P, Herbelin J-M. Corrosion Science 34 (1993) 1123.

[29] Marcus P, Oudar J, Olefjord I, Microsc J. Spectr. Electron. 4 (1979) 63.

[30] Ayoub H, Lair V, Griveau S, Brunswick P, Bedioui F, Cassir M. Electroanalysis 24 (2012) 1324.

Section 2

Developments of Electrochemical Materials and Their Applications

Chapter 4

Cyclohexane-Based Liquid-Biphasic Systems for Organic Electrochemistry

Yohei Okada and Kazuhiro Chiba

Additional information is available at the end of the chapter

http://dx.doi.org/10.5772/53362

1. Introduction

Organic chemistry research comprises three fundamental elements, including synthesis, separation, and analysis. The long and untiring efforts of synthetic chemists have established countless useful reactions to enable the preparation of nearly anything. Beneficial and complex structures can be elaborated from abundant and simple starting materials. For example, several elegant synthetic strategies for Oseltamivir (commonly known as "Tamiflu"), an effective antiviral drug for the flu virus, have been proposed (Fig. 1) [1-6]. With concerns about the environmental aspects of these syntheses, various green processes, e.g., transition-metal-free transformations or the use of water as a reaction solvent, have been studied intensively. In addition, outstanding technological advances have been achieved in the field of analysis, realizing numerous powerful methods. Spectroscopy in particular can yield much information about the structure of both naturally-occurring and artificial compounds.

In these contexts, separation has assumed a key role in organic chemistry. It is generally meaningless if the desired compound cannot be isolated from the reaction mixture even though it was prepared and characterized in a precise manner. Typically, recrystallization and silica gel column chromatography are employed as isolation methodologies in industrial and academic fields. While these separation techniques offer high-performance compound isolation, time-consuming preliminary experiments and the use of large amounts of silica gel are required. Solid-phase techniques are one solution to provide great advantages with respect to compound separation and have also proven to be effective for automated synthesis and combinatorial chemistry. Reaction substrates are generally bound to a solid-phase, enabling their separation from the reaction mixture merely through filtration and washing with solvents. Based on this strategy, efficient multistep chemical transformations, especially for peptide synthesis, can be accomplished. Additionally, the immobilization of reaction catalysts on a solid-phase is efficient for their consecutive recycling, and can also serve as a promising application in combination with a flow strategy.

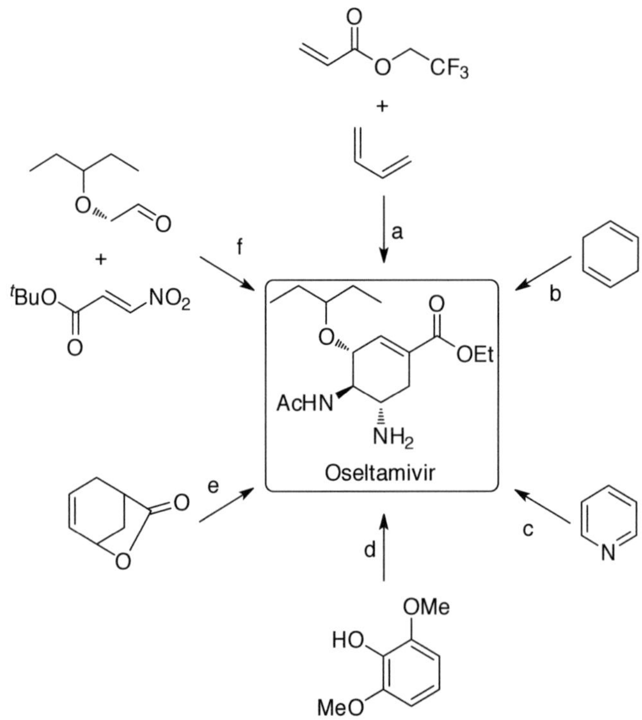

(a) Yeung, Y.-Y.; Hong, S.; Corey, E. J. *J. Am. Chem. Soc.* **2006**, *128*, 6310-6311.
(b) Fukuta, Y.; Mita, T.; Fukuda, N.; Kanai, M.; Shibasaki, M. *J. Am. Chem. Soc.* **2006**, *128*, 6312-6313.
(c) Satoh, N.; Akiba, T.; Yokoshima, S.; Fukuyama, T. *Angew. Chem. Int. Ed.* **2007**, *46*, 5734-5736.
(d) Zutter, U.; Iding, H.; Spurr, P.; Wirz, B. *J. Org. Chem.* **2008**, *73*, 4895-4902.
(e) Trost, B. M.; Zhang, T. *Angew. Chem. Int. Ed.* **2008**, *47*, 3759-3761.
(f) Ishikawa, H.; Suzuki, T.; Hayashi, Y. *Angew. Chem. Int. Ed.* **2009**, *48*, 1304-1307.

Figure 1. Synthetic strategies for Oseltamivir

In addition to the solid-phase technique, the liquid-biphasic technique can also provide a facile separation of compounds by simple liquid-liquid extraction. A representative liquid-biphasic technique is based on the insolubility of perfluorinated hydrocarbons with both polar and less-polar organic solvents, known as fluorous systems [7-11]. In these systems, fluorous compounds, including substrates, products, and catalysts, or designed fluorous platforms, are preferentially dissolved into the fluorous phase to enable their rapid separation. Moreover, "thermomorphic" systems have been developed to offer unique liquid-biphasic separation techniques that change thermally from biphasic conditions to monophasic conditions [12-19].

In organic electrochemistry, electrodes have been utilized as solid-phase redox reagents to trigger either one- or two-electron transfers that afford various functional group transformations and a wide variety of carbon-carbon bond formations in a controlled manner [20-25]. In particular, there is good chemistry between electrochemical approaches and cyclic compounds to produce complex ring systems in one step. For example, five-, six-, and seven-membered rings can be constructed through ring rearrangement or intramolecular cycloaddition (Fig. 2) [26-28]. We have also been developing a series of electrochemical intermolecular cycloadditions initiated by anodic oxidation to give four-, five, and six-membered rings (Fig. 3) [29-31].

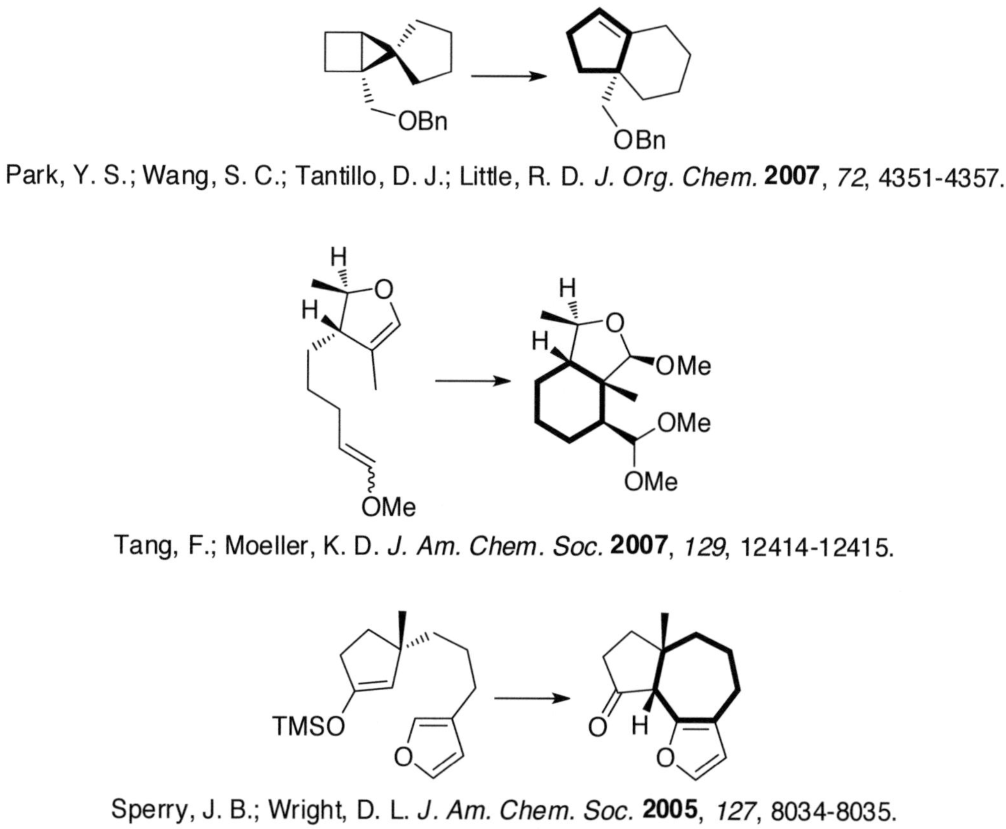

Figure 2. Electrochemical intramolecular five-, six-, and seven-membered ring formations

Although electrodes can be rapidly removed from the reaction mixtures after the completion of electrochemical transformations, the separation of products from supporting electrolytes that are necessary for imparting electrical conductivity to polar organic solvents is still required. In order to address this problem, various ingeniously designed electrochemical reaction systems have been developed [32-36]. In this chapter, we describe cyclohexane-based liquid-biphasic systems as unique separation techniques that are well-combined with organic electrochemistry. The combination of electrodes as solid-phase redox reagents and cyclohexane-based liquid-biphasic systems has paved the way for organic electrochemistry.

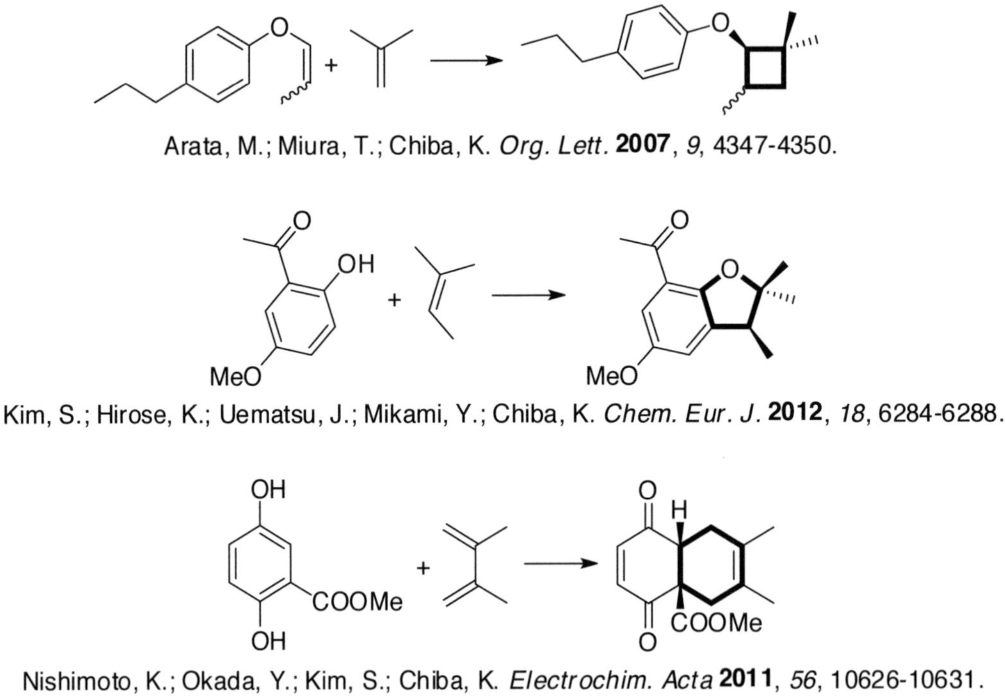

Figure 3. Electrochemical intermolecular four-, five-, and six-membered ring formations

2. Cyclohexane-based liquid-biphasic systems

Cyclohexane-based liquid-biphasic systems have their roots in the initial discovery that cyclohexane has unique thermomorphic properties [37]. Numerous investigations aimed at constructing new liquid-biphasic systems have led to the finding that cyclohexane can be used to successfully form thermomorphic biphasic solutions with typical polar organic solvents and that the regulation of their separation and mixing can be achieved by moderate control in a practical temperature range. A 1:4 (v/v) mixture of cyclohexane and nitromethane, for example, exhibits biphasic conditions at 25 °C, then forms a monophasic condition at *ca.* 60 °C and higher (Fig. 4). In this system, the thermally-mixed monophasic condition can serve as an effective homogeneous reaction field between a less polar substrate that dissolves selectively into the cyclohexane phase and a polar substrate that dissolves selectively into the nitromethane phase. This monophasic solution is cooled to reform the biphasic condition after completion of the reaction, and the hydrophobic products or designed hydrophobic platforms dominantly partition into the cyclohexane phase, which can be recovered rapidly [38-41].

http://dx.doi.org/10.5772/53362

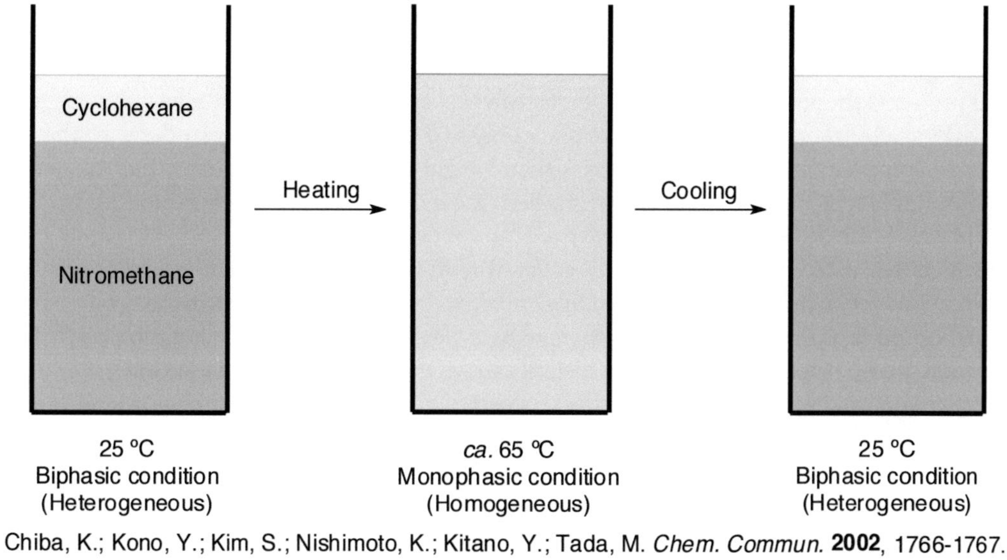

Figure 4. Thermomorphic property of cyclohexane in combination with nitromethane

Moreover, several other polar organic solvents can be introduced into the cyclohexane-based liquid-biphasic system and their phase switching temperatures are tunable based on the choice of polar organic solvents and their ratio to cyclohexane. For instance, a 1:3 (v/v) mixture of cyclohexane and acetonitrile is heated to form a monophasic condition at *ca.* 53 °C and higher, whereas heating to *ca.* 40 °C and higher is enough for a 1:1 (v/v) mixture of cyclohexane and methanol to be thermally mixed into a monophasic condition. In light of these studies, organic electrochemistry will benefit from cyclohexane-based liquid-biphasic systems, especially over the issue of separation.

3. Kolbe-coupling assisted by cyclohexane-based liquid-biphasic systems

In order to apply cyclohexane-based liquid-biphasic techniques to organic electrochemistry, we initially investigated a wide variety of compositions of electrolyte solutions composed of polar organic solvents and supporting electrolytes that showed practical thermomorphic properties in combination with cyclohexane. Through numerous trials, we found that a 1:1:2:4 (v/v/v/v) mixture of pyridine, methanol, acetonitrile, and cyclohexane could be thermally mixed to form a monophasic condition even in the presence of saturated potassium hydroxide as a supporting electrolyte. Heating to *ca.* 48 °C and higher was sufficient to create monophasic conditions, while biphasic conditions were reformed when the solvent mixture was cooled to 25 °C [42]. Furthermore, a 0.10 M concentration of lithium perchlorate could be employed as a supporting electrolyte to show reversible thermal phase switching (Fig. 5).

With these results in hand, Kolbe-coupling, known as a representative electrochemical reaction in organic chemistry, was then carried out in the cyclohexane-based liquid-biphasic system (Fig. 6). Essentially, electrochemical approaches have the requirement that both substrates and products should be soluble in polar electrolyte solutions. This is due to the following reasons. First, electron transfer events take place only at the surface of the electrodes such that insoluble compounds are unable to access their neighborhood, which means that the use of hydrophobic compounds is generally restricted. Second, the formation of insoluble hydrophobic products during the electrochemical transformations might result in electrode passivation in which the surface of the electrodes is covered with polymeric films that severely suppress electric current. In this regard, the thermally-mixed monophasic conditions in the cyclohexane-based liquid-biphasic system can be deemed as a "less-polar" electrolyte solution because it contains an equal volume of less-polar cyclohexane as the polar electrolyte solution.

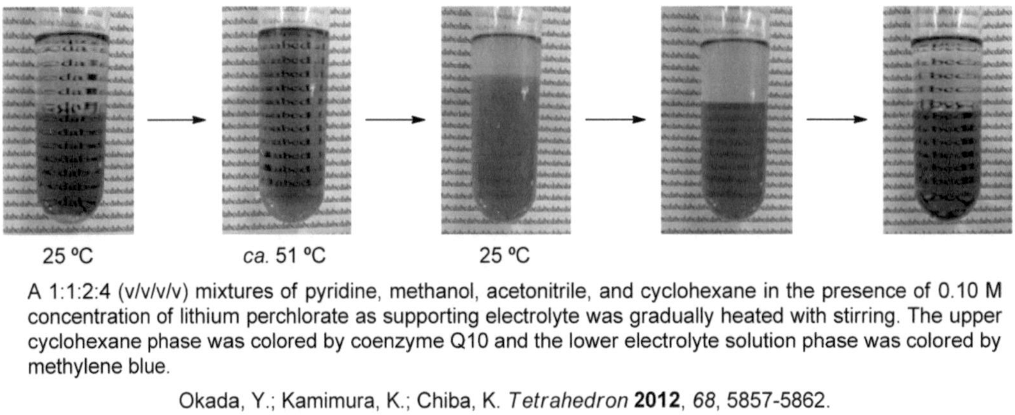

Figure 5. Reversible thermal phase switching of the cyclohexane-based liquid-biphasic system

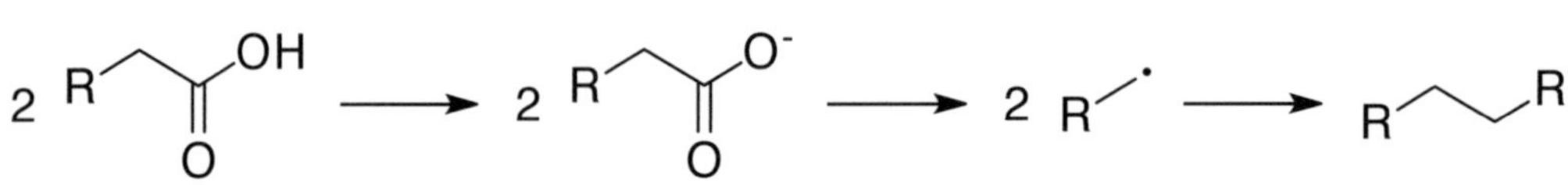

Figure 6. Kolbe-coupling including its proposed reaction mechanism

To test this idea, octanoic acid (**1**) was chosen as a simple model for Kolbe-coupling. Not surprisingly, the reaction proceeded nicely to give the coupled product (**2**) in excellent yield using conventional conditions, i.e., the electrolysis was carried out without cyclohexane (Fig. 7). As expected, the Kolbe-coupling also took place effectively in the cyclohexane-based liquid-biphasic system to suggest its possibility for applications to organic electrochemistry. In this case, the electrochemical reaction was carried out in the thermally-mixed monophasic condition, which was cooled to reform the biphasic condition, realizing rapid separation of the product (**2**) simply through liquid-liquid extraction (Fig. 8). The coupled product (**2**) was so hydrophobic that it was dissolved selectively in the cyclohexane phase.

http://dx.doi.org/10.5772/53362

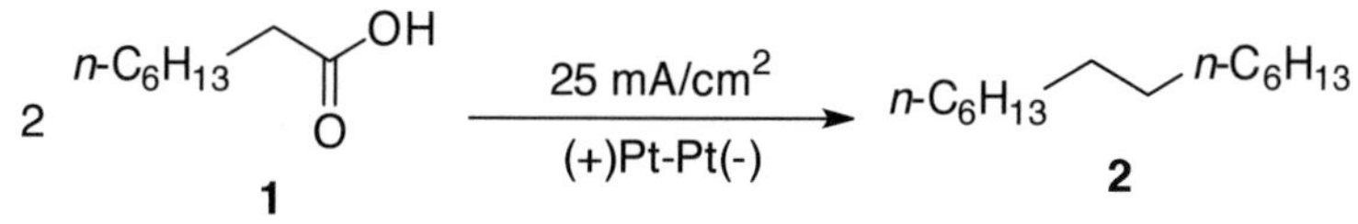

electrolyte solution	yield (%)
pyridine:MeOH:MeCN = 1:1:2	93
pyridine:MeOH:MeCN:*c*-Hex = 1:1:2:4	95

Okada, Y.; Kamimura, K.; Chiba, K. *Tetrahedron* **2012**, *68*, 5857-5862.

Figure 7. Evaluation of the cyclohexane-based liquid-biphasic system for organic electrochemistry

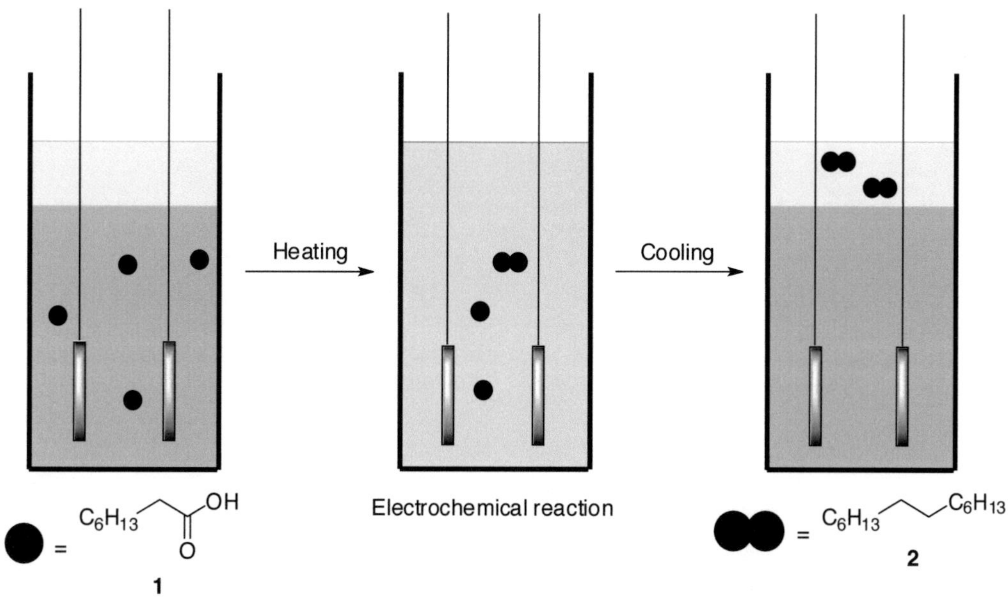

Okada, Y.; Kamimura, K.; Chiba, K. *Tetrahedron* **2012**, *68*, 5857-5862.

Figure 8. Overall reaction procedure of the Kolbe-coupling

Several carboxylic acids (**3-6**) were subsequently used for the Kolbe-coupling in the cyclohexane-based liquid-biphasic system to give the corresponding coupled products (**7-10**) effectively (Fig. 9). These coupled products were poorly soluble in polar electrolyte solution, generally causing electrode passivation to decrease the reaction yield. In contrast, the reaction yields were significantly improved when the cyclohexane-based liquid-biphasic system was used. After the electrochemical reaction, the biphasic condition was reformed through cooling to enable the facile separation of the coupled products (**7-10**). These results show the possibility of using cyclohexane-based liquid-biphasic systems in organic electrochemistry.

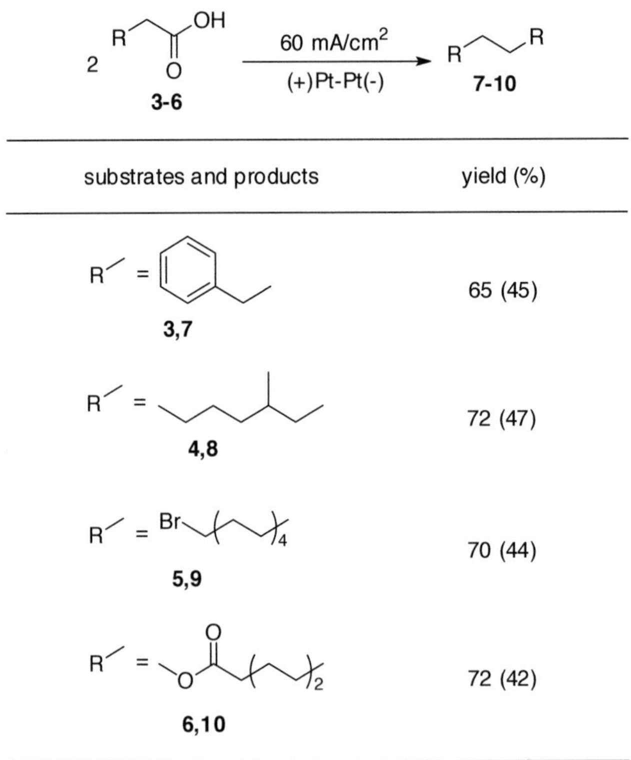

Yields using conventional conditions are shown in parentheses.
Okada, Y.; Kamimura, K.; Chiba, K. *Tetrahedron* **2012**, *68*, 5857-5862.

Figure 9. Kolbe-coupling in the cyclohexane-based liquid-biphasic system

4. Five-membered ring formation assisted by cyclohexane-based liquid-biphasic systems

Undesired overoxidation also becomes problematic in organic electrochemistry. While this is not a concern for the Kolbe-coupling because the oxidation potential of the product is generally lower than that of the substrate, it might severely decrease the reaction yield in some instances. For example, a creative solution to the electrochemical five-membered ring formation between 4-methoxyphenol (**11**) and 2-methylbut-2-ene (**12**) must be developed to avoid this problem (Fig. 10) [43]. The oxidation potential of the five-membered ring product (**13**) is lower than that of 4-methoxyphenol (**11**), thus, undesired overoxidation is possible. As de-

http://dx.doi.org/10.5772/53362

scribed above, the electron transfer events in organic electrochemistry occur only at the surface of the electrodes. Using this situation to its best advantage can be a great aid for problematic overoxidation, namely, if the products can be rapidly removed from the electrodes, their overoxidation could be avoided. For this purpose, the cyclohexane-based liquid-biphasic system is promising because the less-polar cyclohexane phase does not have dissolving power for supporting electrolytes, thus there is no electrical conductivity. In other words, the cyclohexane phase is isolated from the electron transfer events.

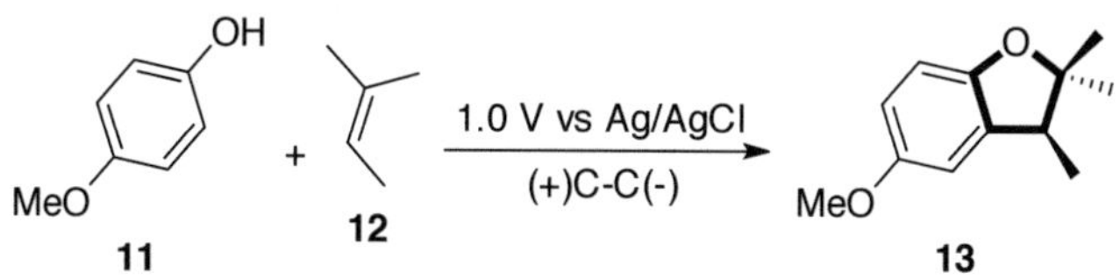

Chiba, K.; Fukuda, M.; Kim, S.; Kitano, Y.; Tada, M. *J. Org. Chem.* **1999**, *64*, 7654-7656.

Figure 10. Electrochemical five-membered ring formation

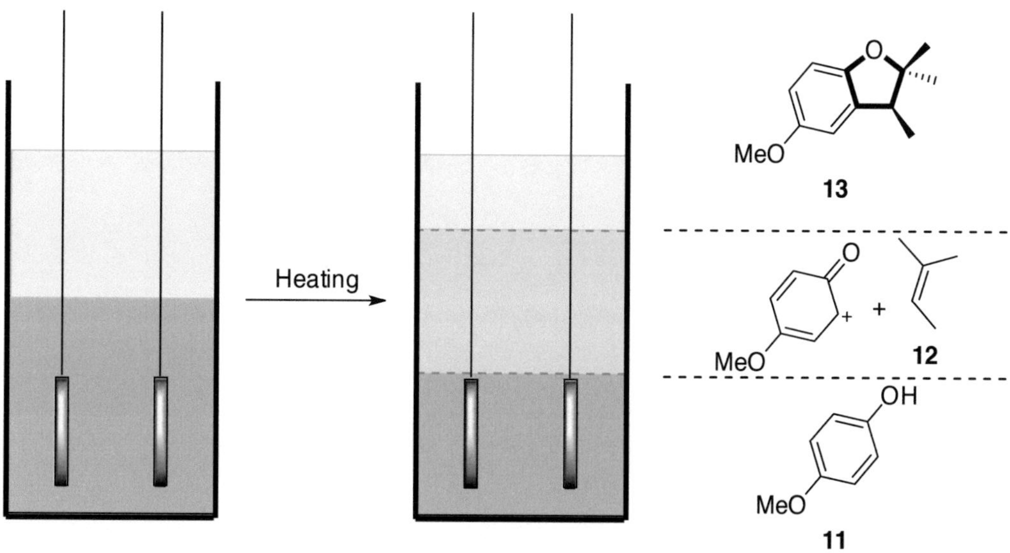

Kim, S.; Noda, S.; Hayashi, K.; Chiba, K. *Org. Lett.* **2008**, *10*, 1827-1830.

Figure 11. Experimental outline of the electrochemical five-membered ring formation

On the basis of this concept, the electrochemical five-membered ring formation between 4-methoxyphenol (**11**) and 2-methylbut-2-ene (**12**) was carried out in the cyclohexane-based liquid-biphasic system (Fig. 11) [44]. This time, the interfacial surface between the cyclohexane phase and the electrolyte solution was partially heated in order to maintain a certain amount of the cyclohexane phase, which was expected to remove the five-membered ring product (**13**) from the electrodes. Less-polar 2-methylbut-2-ene (**12**) was selectively dissolved into the cyclohexane phase, while relatively polar 4-methoxyphenol (**11**) preferred the electrolyte solution phase. The overoxidation of the product (**13**) was effectively avoided to improve the reaction yield significantly (Fig. 12). In addition to 2-methylbut-2-ene (**12**), various olefin nucleophiles (**14-16**) could be introduced into this system to construct the corresponding five-membered ring products (**17-19**), which were recovered into the cyclohexane phase such that their separation required only liquid-liquid extraction. These results have also highlighted the power of cyclohexane-based liquid-biphasic systems in organic electrochemistry. In addition to facile separation of the products, this system protects the products from their overoxidation.

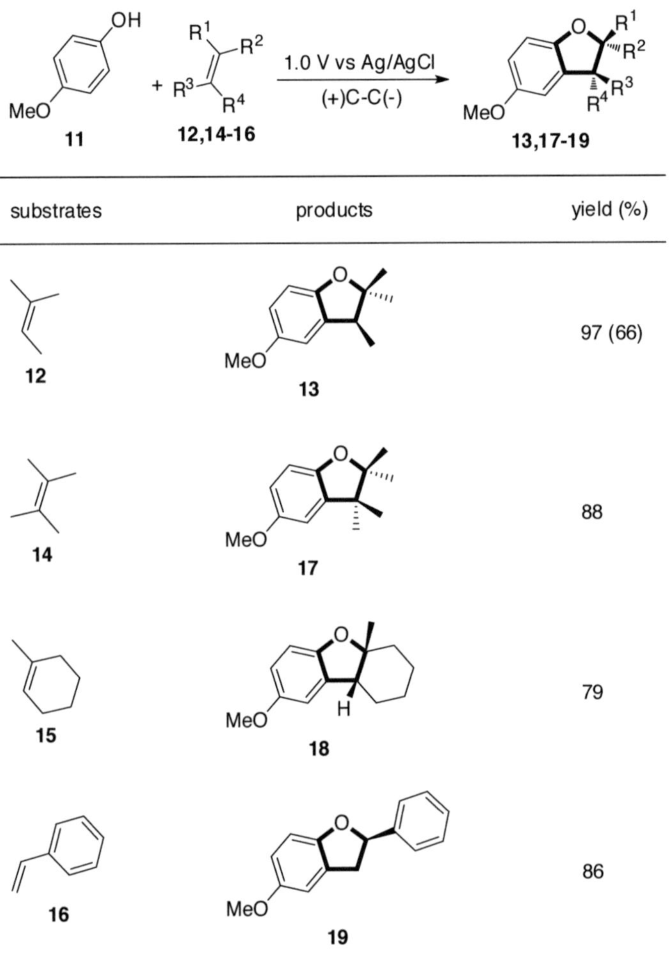

Yield using conventional condition is shown in parenthesis.
Kim, S.; Noda, S.; Hayashi, K.; Chiba, K. *Org. Lett.* **2008**, *10*, 1827-1830.

Figure 12. Electrochemical five-membered ring formations in the cyclohexane-based liquid-biphasic system

5. Four-membered ring formation assisted by cyclohexane-based liquid-biphasic systems

From the environmental viewpoint, the separation process of products and supporting electrolytes is not the only problem in organic electrochemistry. The use of a large amount of supporting electrolytes, which are essential to impart electrical conductivity to polar organic solvents, also causes disposal issues. As described above, although various ingeniously designed electrochemical reaction systems have been reported so far that avoid the use of supporting electrolytes, there are also many electrochemical reactions that are dependent on the presence of a high concentration of supporting electrolytes. We have been developing several electrochemical four-membered ring formations between enol ethers and olefins in nitromethane using high concentrations of lithium perchlorate [45-51]. Because a high concentration of lithium perchlorate in nitromethane can effectively stabilize carbocations and enhance nucleophilicity of olefins, these reactions only take place under such conditions. Therefore, not only organic electrochemistry without supporting electrolytes but also the possibility of their reuse should be considered. For this purpose, cyclohexane-based liquid-biphasic systems are powerful. The less-polar cyclohexane phase does not have dissolving power for supporting electrolytes, thus, they can be confined in the polar electrolyte solution phase, which can be reused for the next reaction (Fig. 13).

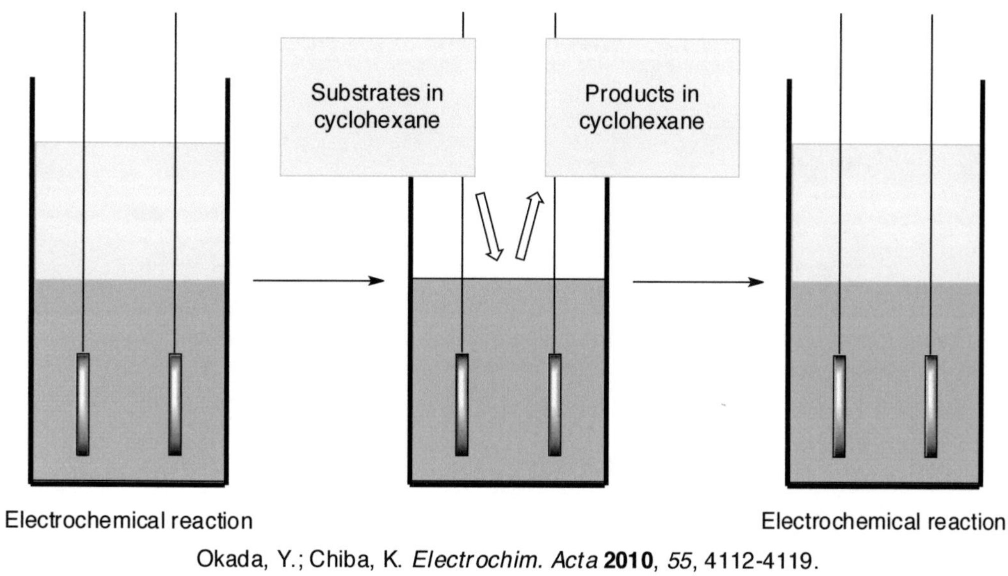

Figure 13. Reuse of supporting electrolytes based on the cyclohexane-based liquid-biphasic system

Based on this concept, the electrochemical four-membered ring formation between 1-(prop-1-en-1-yloxy)-4-propylbenzene (**20**) and hex-1-ene (**21**) was carried out in the cyclohexane-based liquid-biphasic system (Fig. 14) [52]. In this case, the reaction was found to take place even under biphasic conditions without heating to construct the corresponding four-membered ring product (**22**) nicely. The product (**22**) was dominantly partitioned into the cyclohexane phase, where it could be isolated rapidly and purely simply through liquid-liquid extraction. To examine the reusability of the supporting electrolyte, additional amounts of 1-(prop-1-en-1-yloxy)-4-propylbenzene (**20**) and hex-1-ene (**21**) in cyclohexane were introduced to the residual electrolyte solution to find that the yield of the four-membered ring product (**22**) was excellent for at least five cycles. This indicated that the supporting electrolytes could be reused in the cyclohexane-based liquid-biphasic system at least five times in the absence of their degradation, also meaning that the productivity of the product was improved five times. In other words, the required amount of the supporting electrolyte for the reaction was reduced significantly through this system. In addition to hex-1-ene (**21**), several olefin nucleophiles (**23-25**) were then introduced to this system (Fig. 15). This time, too, the four-membered ring product was recovered simply by liquid-liquid extraction at the end of each reaction, and the residual supporting electrolyte was reused five times. Even in this situation, several types of four-membered ring products (**22, 26-28**) could be completely separated from the electrolyte solution to give the corresponding products with high selectivity by simple liquid-liquid extraction. Low productivity, which is one of the weak points of organic electrochemistry, was improved significantly by using the cyclohexane-based liquid-biphasic system.

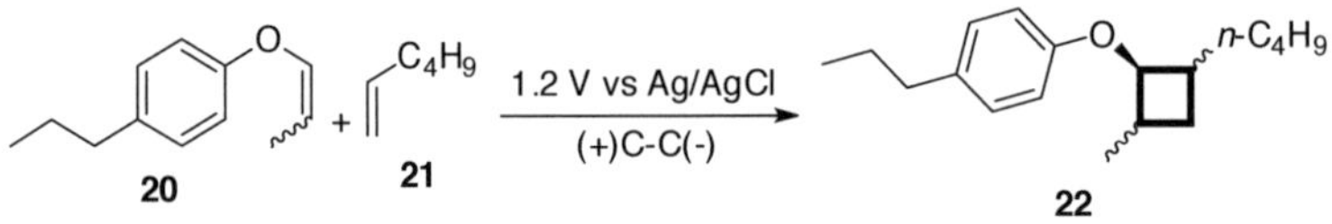

number of trial	1	2	3	4	5
yield (%)	95	95	96	96	95

Okada, Y.; Chiba, K. *Electrochim. Acta* **2010**, *55*, 4112-4119.

Figure 14. Reusability of supporting electrolyte in the cyclohexane-based liquid-biphasic system

6. Continuous flow electrochemical synthesis assisted by cyclohexane-based liquid-biphasic systems

Cyclohexane-based liquid biphasic systems have been well combined with three types of organic electrochemistry, including Kolbe-couplings, five-, and four-membered ring formations, to realize rapid separation of the resulting products. In all cases, the products were selectively dissolved in the cyclohexane phase and could be isolated by simple liquid-liquid extraction to give the desired compounds in almost pure fashion without additional separation steps. Moreover, cyclohexane-based liquid-biphasic systems also offer many valuable

http://dx.doi.org/10.5772/53362

functions as follows. In Kolbe-couplings, the biphasic condition was thermally mixed to a monophasic condition to form a less-polar electrolyte solution, which avoided electrode passivation effectively to improve the reaction efficiency. In electrochemical five-membered ring formations, the cyclohexane phase selectively dissolved the products to protect them from undesired overoxidation. High reusability of the supporting electrolyte was also demonstrated and improvement of productivity was achieved in electrochemical four-membered ring formations. Here we have set an ultimate aim to accomplish flow electrochemical synthesis assisted by cyclohexane-based liquid-biphasic systems.

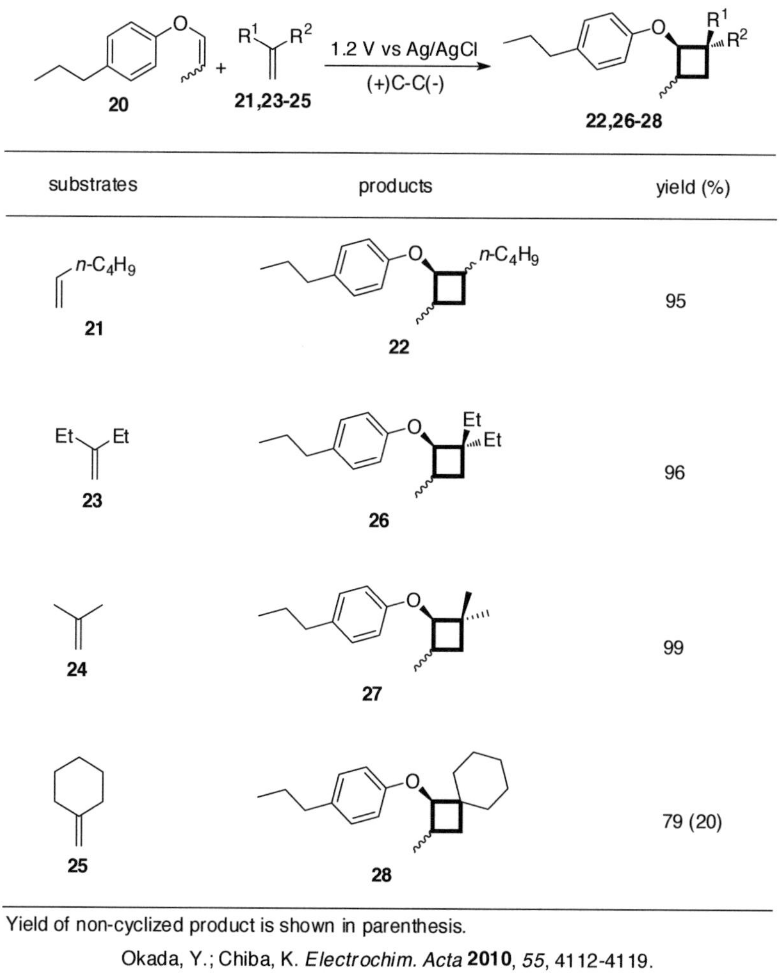

Figure 15. Electrochemical four-membered ring formations in the cyclohexane-based liquid-biphasic system

For the construction of a flow electrochemical synthetic device, the composition of several cyclohexane-based liquid-biphasic systems was studied in detail (Fig. 16). Three electrolyte solutions were prepared using 1.0 M lithium perchlorate as a supporting electrolyte in methanol, acetonitrile, or nitromethane. The same volume of cyclohexane was then added to investigate the compositions of both upper and lower phases.

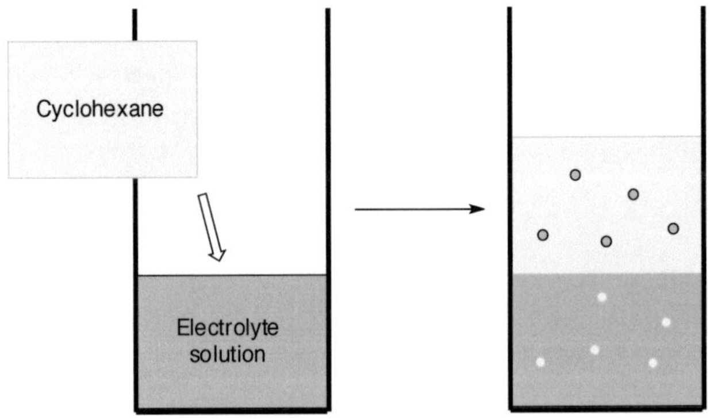

ratio	compositions (upper phase)	compositions (lower phase)
c-Hex:MeOH = 1:1	*c*-Hex:MeOH = 96:4 (<0.3 mM)	*c*-Hex:MeOH = 5:95 (*ca.* 1.0 M)
c-Hex:MeCN = 1:1	*c*-Hex:MeCN = 97:3 (<0.4 mM)	*c*-Hex:MeCN = 3:97 (*ca.* 1.0 M)
c-Hex:$MeNO_2$ = 1:1	*c*-Hex:$MeNO_2$ = 98:2 (<0.5 mM)	*c*-Hex:$MeNO_2$ = 2:98 (*ca.* 1.0 M)

Concentrations of $LiClO_4$ are shown in parenthesis.

Okada, Y.; Yoshioka, T.; Koike, M.; Chiba, K. *Tetrahedron Lett.* **2011**, *52*, 4690-4693.

Figure 16. Compositions of cyclohexane-based liquid-biphasic systems

Remarkably, although cyclohexane and these polar solvents were partially miscible even at ambient temperature, only a trace amount of lithium perchlorate was recovered from the cyclohexane phase. This meant that the supporting electrolyte could be confined in polar solvents. As described, electrodes function as solid-phase redox reagents; therefore, they should be well combined with a flow strategy (Fig. 17) [53]. Substrates in cyclohexane are injected into an electrochemical reactor, which is equipped with electrodes and a filter that is selectively permeable to cyclohexane. In this system, the outlet ejected from the reactor might be almost pure product in cyclohexane.

Based on these preliminary experiments, we designed and prepared a new flow electrochemical synthetic device (Fig. 18). The device was built with three compartments, which are mainly made of polytetrafluoroethylene, and could contain 5.0 mL of electrolyte solution. The third compartment was filled with polytetrafluoroethylene fiber that is known to be permeable for less-polar cyclohexane rather than polar electrolyte solutions. Substrates could be pumped into the reactor as a cyclohexane solution from the inlet, which would then emerge from the outlet after the electrochemical reaction.

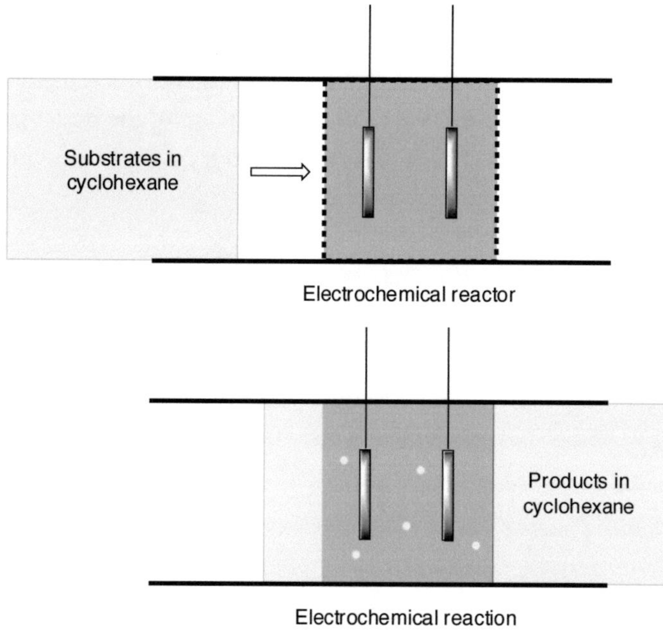

Okada, Y.; Yoshioka, T.; Koike, M.; Chiba, K. *Tetrahedron Lett.* **2011**, *52*, 4690-4693.

Figure 17. Outline of flow electrochemical synthesis based on the cyclohexane-based liquid-biphasic system

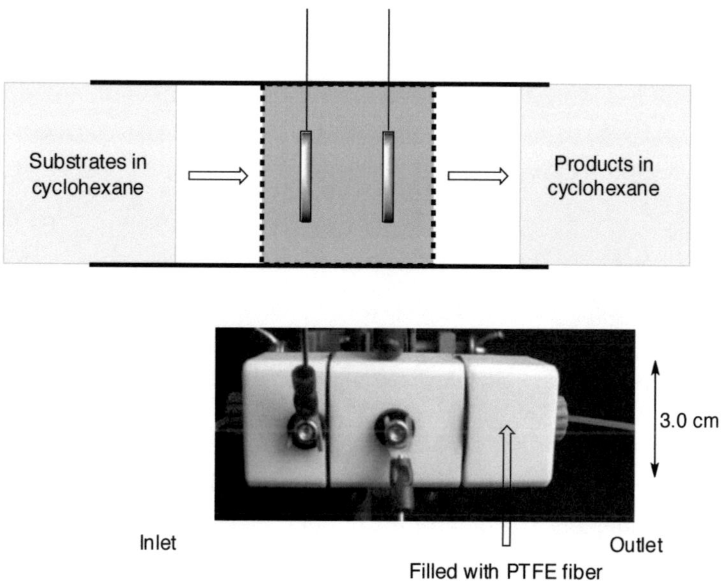

Okada, Y.; Yoshioka, T.; Koike, M.; Chiba, K. *Tetrahedron Lett.* **2011**, *52*, 4690-4693.

Figure 18. Design of new flow electrochemical synthetic device

With this device in hand, initially, electrochemical methoxylation, another representative electrochemical reaction in organic chemistry, was attempted (Fig. 19). The methoxylation of hydrophobic furan (**29**) nicely took place in the reactor to afford the methoxylated product (**30**) in a selective maanner (Fig. 20). Vacuum concentration of the outlet cyclohexane solution ejected from the reactor was all that was required to give nearly pure product.

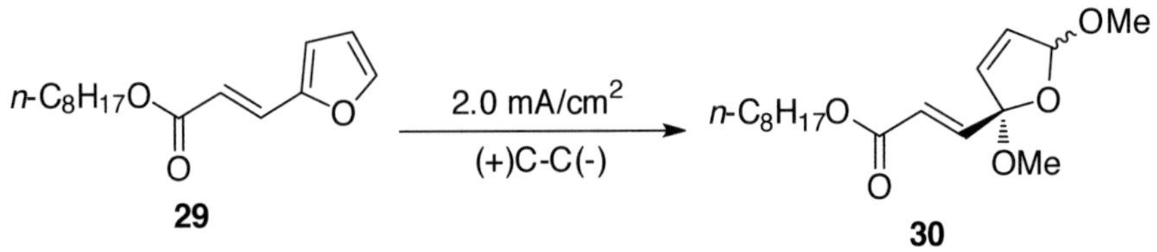

Figure 19. Electrochemical methoxylation

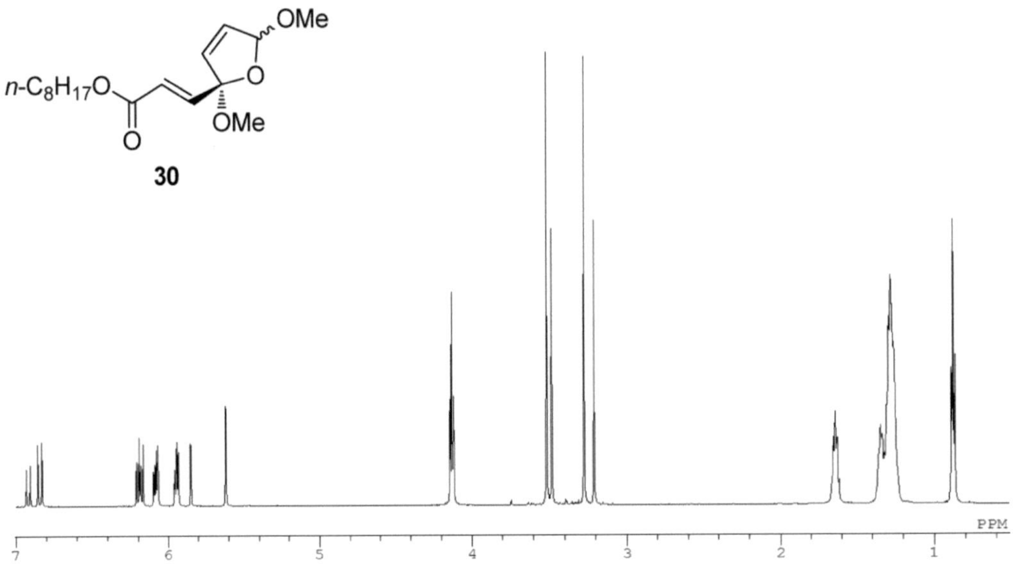

Figure 20. NMR spectrum of the methoxylated product (**30**)

Finally, electrochemical four-, five-, and six-membered ring formations were carried out in this flow system (Fig. 21). All reactions took place selectively in the reactor and it should be noted that only vacuum concentration of the outlet cyclohexane solution was required. The corresponding ring products could be obtained in almost pure fashion without any additional separation processes. It is perhaps fair to say that the desired products emerge automatically from the flow electrochemical synthetic device assisted by the cyclohexane-based liquid-biphasic system.

http://dx.doi.org/10.5772/53362

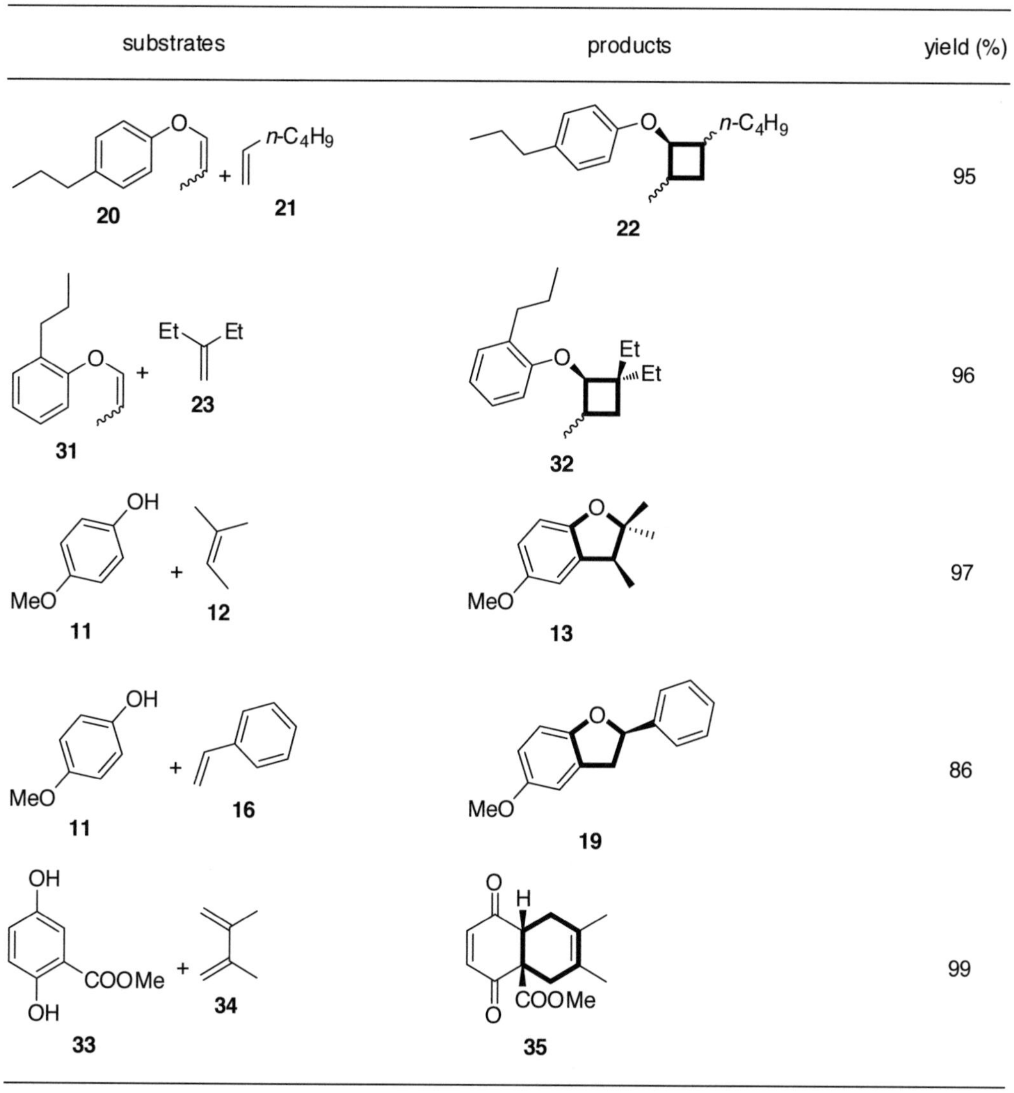

Okada, Y.; Yoshioka, T.; Koike, M.; Chiba, K. *Tetrahedron Lett.* **2011**, *52*, 4690-4693.

Figure 21. Flow electrochemical synthesis assisted by the cyclohexane-based liquid-biphasic system

7. Conclusion

As described in this chapter, the discovery that cyclohexane had a unique thermomorphic nature has led to the development of cyclohexane-based liquid-biphasic systems, which can be well-combined with organic electrochemistry. Rapid and high-performance separations, which have been assuming a larger role in modern organic chemistry, were accomplished by this system. Cyclohexane-based liquid-biphasic systems offer not only effective separation but also several additional functions of value, including suppressing electrode passivation, protecting products from overoxidation, and enabling reuse of supporting electrolytes.

Desired compounds can be synthesized precisely, and then characterized carefully. Organic chemists have made significant advances in these techniques to realize the preparation of almost anything with detailed structural information. In this context, the development of effective separation methodologies should maximize their vitality. For this purpose, the cyclohexane-based biphasic-system is one of the most promising techniques, especially in organic electrochemistry.

Acknowledgments

This work was partially supported by a Grant-in-Aid for Scientific Research from the Ministry of Education, Culture, Sports, Science, and Technology.

Author details

Yohei Okada and Kazuhiro Chiba

Tokyo University of Agriculture and Technology, Japan

References

[1] Yeung Y-Y, Hong S, Corey EJ. A Short Enantioselective Pathway for the Synthesis of the Anti-Influenza Neuramidase Inhibitor Oseltamivir from 1,3-Butadiene and Acrylic Acid. Journal of the American Chemical Society 2006;128(19) 6310-6311.

[2] Fukuta Y, Mita T, Fukuda N, Kanai M, Shibasaki M. De Novo Synthesis of Tamiflu via a Catalytic Asymmetric Ring-Opening of meso-Aziridines with $TMSN_3$. Journal of the American Chemical Society 2006;128(19) 6312-6313.

[3] Satoh N, Akiba T, Yokoshima S, Fukuyama T. A Practical Synthesis of (−)-Oseltamivir. Angewandte Chemie International Edition 2007;46(30) 5734-5736.

[4] Zutter U, Iding H, Spurr P, Wirz B. New, Efficient Synthesis of Oseltamivir Phosphate (Tamiflu) via Enzymatic Desymmetrization of a meso-1,3-Cyclohexanedicarboxylic Acid Diester. The Journal of Organic Chemistry 2008;73(13) 4895-4902.

[5] Trost BM, Zhang T. A Concise Synthesis of (−)-Oseltamivir. Angewandte Chemie International Edition 2008;47(20) 3759-3761.

[6] Ishikawa H, Suzuki T, Hayashi Y. High-Yielding Synthesis of the Anti-Influenza Neuramidase Inhibitor (−)-Oseltamivir by Three "One-Pot" Operations. Angewandte Chemie International Edition 2009;48(7) 1304-1307.

[7] Zhang W. Fluorous Linker-Facilitated Chemical Synthesis. Chemical Reviews 2009;109(2) 749-795.

[8] Zhang W. Fluorous Synthesis of Heterocyclic Systems. Chemical Reviews 2004;104(5) 2531-2556.

[9] Studer A, Hadida S, Ferritto R, Kim S-Y, Jeger P, Wipf P, Curran DP. Fluorous Synthesis: A Fluorous-Phase Strategy for Improving Separation Efficiency in Organic Synthesis. Science 1997;275(5301) 823-826.

[10] Horvath IT, Rabai J. Facile Catalyst Separation Without Water: Fluorous Biphase Hydroformylation of Olefins. Science 1994;266(5182) 72-75.

[11] Gladysz JA. Are Teflon "Ponytails" the Coming Fashion for Catalysts? Science 1994;266(5182) 55-56.

[12] Bergbreiter DE, Tian J, Hongfa C. Using Soluble Polymer Supports To Facilitate Homogeneous Catalysis. Chemical Reviews 2009;109(2) 530-582.

[13] Bergbreiter DE. Using Soluble Polymers To Recover Catalysts and Ligands. Chemical Reviews 2002;102(10) 3345-3384.

[14] Hobbs C, Yang Y-C, Ling J, Nicola S, Su H-L, Bazzi HS, Bergbreiter DE. Thermomorphic Polyethylene-Supported Olefin Metathesis Catalysts. Organic Letters 2011;13(15) 3904-3907.

[15] Bergbreiter DE, Liu Y-S, Osburn PL. Thermomorphic Rhodium(I) and Palladium(0) Catalysts. Journal of the American Chemical Society 1998;120(17) 4250-4251.

[16] Huang Y-Y, He Y-M, Zhou H-F, Wu L, Li B-L, Fan Q-H. Thermomorphic System with Non-Fluorous Phase-Tagged Ru(BINAP) Catalyst: Facile Liquid/Solid Catalyst Separation and Application in Asymmetric Hydrogenation. The Journal of Organic Chemistry 2006;71(7) 2874-2877.

[17] Barré G, Taton D, Lastécouères D, Vincent J-M. Closer to the "Ideal Recoverable Catalyst" for Atom Transfer Radical Polymerization Using a Molecular Non-Fluorous Thermomorphic System. Journal of the American Chemical Society 2004;126(25) 7764-7765.

[18] Rocaboy C, Gladysz JA. Highly Active Thermomorphic Fluorous Palladacycle Catalyst Precursors for the Heck Reaction; Evidence for a Palladium Nanoparticle Pathway. Organic Letters 2002;4(12) 1993-1996.

[19] Wende M, Meier R, Gladysz JA. Fluorous Catalysis without Fluorous Solvents: A Friendlier Catalyst Recovery/Recycling Protocol Based upon Thermomorphic Properties and Liquid/Solid Phase Separation. Journal of the American Chemical Society 2001;123(46) 11490-11491.

[20] Yoshida J, Kataoka K, Horcajada R, Nagaki, A. Modern Strategies in Electroorganic Synthesis. Chemical Reviews 2008;108(7) 2265-2299.

[21] Sperry JB, Wright DL. The application of cathodic reductions and anodic oxidations in the synthesis of complex molecules. Chemical Society Reviews 2006;35(7) 605-621.

[22] Little RD. Diyl Trapping and Electroreductive Cyclization Reactions. Chemical Reviews 1996;96(1) 93-114.

[23] Moeller KD. Intramolecular Anodic Olefin Coupling Reactions: Using Radical Cation Intermediates to Trigger New Umpolung Reactions. Synlett 2009;(8) 1208-1218.

[24] Moeller KD. Synthetic Applications of Anodic Electrochemistry. Tetrahedron 2000;56(49) 9527-9554.

[25] Fuchigami T, Inagi S. Selective Electrochemical Fluorination of Organic Molecules and Macromolecules in Ionic Liquids. Chemical Communications 2011;47(37) 10211-10223.

[26] Park YS, Wang SC, Tantillo DJ, Little RD. A Highly Selective Rearrangement of a Housane-Derived Cation Radical: An Electrochemically Mediated Transformation. The Journal of Organic Chemistry 2007;72(12) 4351-4357.

[27] Tang F, Moeller KD. Intramolecular Anodic Olefin Coupling Reactions: The Effect of Polarization on Carbon–Carbon Bond Formation. Journal of the American Chemical Society 2007;129(41) 12414-12415.

[28] Sperry JB, Wright DL. The gem-Dialkyl Effect in Electron Transfer Reactions: Rapid Synthesis of Seven-Membered Rings through an Electrochemical Annulation. Journal of the American Chemical Society 2005;127(22) 8034-8035.

[29] Arata M, Miura, T, Chiba, K. Electrocatalytic Formal [2+2] Cycloaddition Reactions between Anodically Activated Enyloxy Benzene and Alkenes. Organic Letters 2007;9(21) 4347-4350.

[30] Kim S, Hirose K, Uematsu J, Mikami Y, Chiba K. Electrochemically Active Cross-Linking Reaction for Fluorescent Labeling of Aliphatic Alkenes. Chemistry – A European Journal 2012;18(20) 6284-6288.

[31] Nishimoto K, Okada Y, Kim S, Chiba K. Rate acceleration of Diels-Alder reactions utilizing a fluorous micellar system in water. Electrochimica Acta 2011;56(28) 10626-10631.

[32] Horcajada R, Okajima M, Suga S, Yoshida J. Microflow electroorganic synthesis without supporting electrolyte. Chemical Communications 2005;(10) 1303-1304.

[33] Sawamura T, Inagi S, Fuchigami T. Use of Task-Specific Ionic Liquid for Selective Electrocatalytic Fluorination. Organic Letters 2010;12(3) 644-646.

[34] Sunaga T, Atobe M, Inagi S, Fuchigami T. Highly efficient and selective electrochemical fluorination of organosulfur compounds in Et_3N 3HF ionic liquid under ultrasonicat. Chemical Communications 2009;(8) 956-958.

[35] Tajima T, Nakajima A. Direct Oxidative Cyanation Based on the Concept of Site Isolation. Journal of the American Chemical Society 2008;130(32) 10496-10497.

[36] Tajima T, Nakajima A, Doi Y, Fuchigami T. Anodic Fluorination Based on Cation Exchange between Alkali-Metal Fluorides and Solid-Supported Acids. Angewandte Chemie International Edition 2007;46(19) 3550-3552.

[37] Chiba K, Kono Y, Kim S, Nishimoto K, Kitano Y, Tada M. A liquid-phase peptide synthesis in cyclohexane-based biphasic thermomorphic systems. Chemical Communications 2002;(16) 1766-1767.

[38] Kim S, Ikuhisa N, Chiba K. A Cycloalkane-based Thermomorphic System for Organocatalytic Cyclopropanation Using Ammonium Ylides. Chemistry Letters 2011;40(10) 1077-1078.

[39] Kim S, Tsuruyama A, Ohmori A, Chiba K. Solution-phase oligosaccharide synthesis in a cycloalkane-based thermomorphic system. Chemical Communications 2008;(15) 1816-1818.

[40] Kim S, Yamamoto K, Hayashi K, Chiba K. A cycloalkane-based thermomorphic system for palladium-catalyzed cross-coupling reactions. Tetrahedron 2008;64(12) 2855-2863.

[41] Hayashi K, Kim S, Kono Y, Tamura M, Chiba K. Microwave-promoted Suzuki-Miyaura coupling reactions in a cycloalkane-based thermomorphic biphasic system Tetrahedron Letters 2006;47(2) 171-174.

[42] Okada Y, Kamimura K, Chiba K. Cycloalkane-based thermomorphic systems for organic electrochemistry: an application to Kolbe-coupling. Tetrahedron 2012;68(29) 5857-5862.

[43] Chiba K, Fukuda M, Kim S, Kitano Y, Tada, M. Dihydrobenzofuran Synthesis by an Anodic [3 + 2] Cycloaddition of Phenols and Unactivated Alkenes. The Journal of Organic Chemistry 1999;64(20) 7654-7656.

[44] Kim S, Noda S, Hayashi K, Chiba K. An Oxidative Carbon–Carbon Bond Formation System in Cycloalkane-Based Thermomorphic Multiphase Solution. Organic Letters 2008;10(9) 1827-1829.

[45] Chiba K, Okada Y. Electron-Transfer-Induced Intermolecular [2 + 2] Cycloaddition Reactions Assisted by Aromatic "Redox Tag". In: Sur UK. (ed.) Recent Trend in Electrochemical Science and Technology. Rijeka: InTech; 2011. p91-106.

[46] Okada Y, Yamaguchi Y, Chiba K. Efficient Intermolecular Carbon–Carbon Bond-Formation Reactions Assisted by Surface-Condensed Electrodes. European Journal of Organic Chemistry 2012;(2) 243–246.

[47] Okada Y, Nishimoto A, Akaba R, Chiba K. Electron-Transfer-Induced Intermolecular [2 + 2] Cycloaddition Reactions Based on the Aromatic "Redox Tag" Strategy. The Journal of Organic Chemistry 2011;76(9) 3470–3476.

[48] Okada Y, Chiba K. Electron transfer-induced four-membered cyclic intermediate formation: Olefin cross-coupling vs. olefin cross-metathesis. Electrochimica Acta 2011;56(3) 1037–1042.

[49] Okada Y, Akaba R, Chiba K. EC-backward-E electrochemistry supported by an alkoxyphenyl group. Tetrahedron Letters 2009;50(38) 5413–5416.

[50] Okada Y, Akaba R, Chiba K. Electrocatalytic Formal [2+2] Cycloaddition Reactions between Anodically Activated Aliphatic Enol Ethers and Unactivated Olefins Possessing an Alkoxyphenyl Group. Organic Letters 2009;11(4) 1033–1035.

[51] Chiba K, Miura T, Kim S, Kitano Y, Tada M. Journal of the American Chemical Society 2001;123(45) 11314-11315.

[52] Okada Y, Chiba K. Continuous electrochemical synthetic system using a multiphase electrolyte solution. Electrochimica Acta 2010;55(13) 4112–4119.

[53] Okada Y, Yoshioka T, Koike M, Chiba K. Heterogeneous continuous flow synthetic system using cyclohexane-based multiphase electrolyte solutions. Tetrahedron Letters 2011;52(36) 4690-4693.

Chapter 5

Electrochemical Transformation of White Phosphorus as a Way to Compounds With Phosphorus-Hydrogen and Phosphorus-Carbon Bonds

Yu. G. Budnikova and S. A. Krasnov

Additional information is available at the end of the chapter

http://dx.doi.org/10.5772/53886

1. Introduction

Organophosphorus compounds (OPC) have gained huge importance in modern manufacturing of bulk and fine chemicals and, apart the classic applications as fertilizers, detergents and pesticides, they represent the basic compounds for the development of intriguing materials for micro- and optoelectronics, coherent and nonlinear optics, selective extractors for rare-earths and transuranic elements, additives for plastic materials (flame retardants, plasticizers, softeners, etc.), additives to lubricant oils and liquid fuels, flotation agents, emulsifiers, etc. More recent OPC applications have come out in the field of biologically active compounds for medicine (chemotherapics, antiviral agents, biocompatible materials for bone and dental reparations, etc.) and in agrochemistry.

The interest in the direct synthesis of OPC from elemental phosphorus, escaping the traditional stages of its chlorination, stems from the increasing needs to reorient the chemical technology towards the strategic target of combining increased ecological safety with low-waste production. In this context, a replacement of the current process for the preparation of basic OPC using phosphorus chloride and oxychloride is of paramount importance. The existing processes are environmentally dangerous, energy-consuming and utterly remote from the concepts marking out green chemistry technological processes. The formation of hydrogen chloride accounting, by mass, for three quarters of the used PCl_3 is the key disadvantage in the P_4 chlorination process.

Until the middle of eighties, electrochemical reactions played a minor role in industrial synthetic chemistry in spite the use of electrochemistry had a long tradition in organic chemistry, as shown for example by the Kolbe reaction to produce symmetrical hydrocarbon dimers by electrochemical oxidative decarboxylation of carboxylic acid salts [1,2]. During the sixties electrochemical methods were used to produce tetraethyl lead and adiponitrile, but such applications in large commodities manufacturing may be considered rather atypical. In spite of the shortage of industrial applications, electrochemical reactions have a number of advantages, such as the generally mild and controllable working conditions and the easy control of process rate. Additionally, the selectivity of the process may be easily controlled by a judicious choice of parameters, such as the density current and the potential [3-6].

In this short report we intend reviewing the use of electrochemical methods to generate OPC from the direct transformation of elemental phosphorus. Particular attention will be paid to illustrate the electrocatalytic processes where the use of a metal catalyst would allow for the highly desirable functionalization of white phosphorus. Homogeneous reactions catalyzed by transition metals proceed through reaction cycles involving the metal in different oxidation states readily detectable by a variety of electrochemical methods (voltammetry, amperometry, chronoamperometry, etc.) [3-7].

2. Experimental

Preparative electrolyses were performed using the direct current source B5-49 in thermostatically controlled cylindrical divided electrolyser (a three-electrode cell) with 40 cm^3 volume. Silver $Ag/AgNO_3$ electrode (10 mM solution in MeCN) served as a reference one. Lead cylinder with surface area of 20 cm^2 was used as a cathode. During electrolysis the electrolyte was stirred with a magnetic stirrer. The aqueous acetate buffer was used as catholyte. The saturated potassium acetate aqueous solution was used as anolyte, and the ceramic membrane divided cathodic and anodic spaces. Platinum wire served as an anode. All manipulations and reactions were carried out under dry Ar atmosphere.

The images of the films were obtained on a HITACHI TM-1000 scanning electron microscope.

NMR-spectroscopy of 1H, 31P, IR-spectroscopy, chromatographic analysis, elemental analysis were used to establish the structure and yields of electrosynthesis products. The obtained physical characteristics of the products correspond to the data available in literature.

The NMR 31P spectra were recorded using CXP-100 Brucker spectrometer (85% H3PO4 as an external standard). The NMR 1H spectra were recorded using MSL-400 Brucker spectrometer.

Alkenes were used without any purification (Acros).

2.1. Electrochemical measurements

A cyclic voltammetry (CVA) was carried out with potentiostat E2P Epsilon (BASi, USA) consisting of detector, personal computer Dell Optiplex 320 with software EpsilonEC-USB-

V200, and electrochemical cell C3 having three-electrode scheme. A linear potential scanning speed was 100 mV/s. Stationary glassy carbon electrode (d = 3.0 mm), platinum electrode (d = 1.5 mm), and DSA (Dimensionally Stable Anode) electrode (d = 1.0 mm) were used as working electrodes. Saturated calomel electrode (SCE) was used as reference electrode in voltammetric measurements. Platinum wire (d = 0.5 mm) was used as auxiliary electrode. Tetrabutylammonium tetrafluorborate with concentration 0.1 M was used as a background salt in voltammetric studies. The method of cyclic voltammetry was used to evaluate of the reduction potentials of alkenes. The tetrabutylammonium tetrafluoroborate (Bu_4NBF_4) (0.1 M) solution in dymethylformamide (DMF) served as electrolyte. The lead electrode (S = 3.14 mm^2) was used as the working electrode. Saturated calomel electrode served as a reference one, Pt wire being the auxiliary electrode.

Electron spin resonance (ESR) measurements were carried out using the program-apparatus complex [8] mounted on the basis of an analog-electrochemical setup with a PI-50-1 potentiostat, a Pr-8 programmer, an X-range ESR spectrometer (Radiopan), and E14-440 analog-to-digital and digital-to-analog converter (L-Card), and a computer.

The WinSim 0.96 program (NIEHS) was used for ESR spectra processing.

2.2. Preparation of solutions

Benzene was dehydrated by distillation over sodium. α-phenyl-*N*-*tert*-butylnitrone (1) and Et_4NBF_4 (Fluka) were used without additional purification. Acetonitrile was purified by triple distillation over $KMnO_4$ and P_2O_5, and Et_4NBF_4 was dried in vacuo for 2 days at 100 °C. Dimethylformamide was distilled, kept for 12 h over calcined K_2CO_3, then distilled repeatedly over CaH_2, and stored over molecular sieves 3A calcined at 300 °C.

A solution of white phosphorus in benzene (~16 mM) was purged with helium through a capillary immersed to the bottom of the cell. The material of the working and auxiliary electrodes was platinum, and $Ag/AgNO_3$ (10 mM) was a reference electrode.

The saturated aqueous potassium acetate solution was prepared by addition of 56 g (1 M) KOH to the stirred solution of 60 g (1 M) acetic acid in 20 ml H_2O and further solvation of the formed precipitate in minimal quantity of distilled water. The acetate buffer solution was prepared using 30 ml acetic acid and 10 g of KOH in 30 ml of H_2O.

2.3. Electrochemical reduction of alkenes in the presence of white phosphorus

The solution for electrolysis was prepared by mixing 16 mM (0.5 g) of white phosphorus, 48 mM of appropriate alkene and 20 ml of aqueous acetic buffer solution, as supporting electrolyte, in 20 ml of H_2O. White phosphorus was emulsified under argon when heated to 50 °C before the initiation of electrolysis. The electrolysis was carried out at 10-20 mA cm^{-2} cathodic current density in the galvanostatic mode (about -2.00 V). The amount of electricity passed through the electrolyte was 3*e* per atom of phosphorus. After completing the electrolysis the organic phase was isolated, washed by water and the residuary initial alkene was evaporated in vacuum. Gaseous phosphine in trace amount was directed to exhaust tube and captured in the adjacent vessel filled with a 1.5% aqueous solution of $HgCl_2$. Inorganic hypophosphorus acid was product in an aqueous part of electrolyte (δP 14 ppm, $J_{H\text{-}P\text{-}H}$ = 580 Hz).

The resulting colorless oil was analyzed by NMR and IR spectroscopy.

1. C_6H_5-CH_2-CH_2-PH_2. Yield: 0.69 g, 31% on phosphorus;

^{31}P-NMR ($CDCl_3$): -141.5 ppm ($J_{H\text{-}P\text{-}H}$ = 193.9 Hz); (lit.: –139.7 ppm (198 Hz) [9]);

IR, cm^{-1}: 2281 (P-H); (lit.: 2280 [9]);

n_d^{20} = 1.5568; (lit.: 1.5532 [2]; 1.5494 [10]);

b.p.: 72ºC (7 Torr); (lit.: 46-48ºC (1 Torr) [10]; 75ºC (8 Torr) [11]);

Anal. Found:, %: C, 69.90; H, 8.67; P, 21.43;

Calc. for $C_8H_{11}P_1$, %: C, 69.55; H, 8.03; P, 22.42;

2. $C_6H_{13}PH_2$. Yield: 0.45 g, 23% on phosphorus;

^{31}P-NMR ($CDCl_3$): -140.8 ppm ($J_{H\text{-}P\text{-}H}$ = 190.0 Hz);

IR, cm^{-1}: 2284 (P-H);

b.p.: 127ºC; (lit.: 127.5-128ºC [12]);

Anal. Found:, %: C, 60.96; H, 12.91; P, 26.13;

Calc. for $C_6H_{15}P_1$, %: C, 61.02; H, 12.71; P, 26.27;

3. C_6H_5-$CH(CH_3)$-CH_2-PH_2. Yield: 1.12 g, 46% on phosphorus;

^{31}P-NMR ($CDCl_3$): -148.7 ppm ($J_{H\text{-}P\text{-}H}$ = 190.3 Hz); (lit.: –148.6 ppm (198 Hz) [9]);

IR, cm^{-1}: 2278 (P-H); (lit.: 2280 [9]);

n_d^{20} = 1.5498; (lit.: 1.5482 [9]);

b.p.: 83-85ºC (7 Torr); (lit.: 65-67ºC (2 Torr) [9]);

Anal. Found:, %: C, 70.89; H, 8.73; P, 20.38;

Calc. for $C_8H_{11}P_1$, %: C, 71.04; H, 8.61; P, 20.35;

4. CH_3-C(O)-O-CH_2-CH_2- PH_2. Yield: 0.43 g, 22 % on phosphorus;

^{31}P-NMR ($CDCl_3$): -155.6 ppm ($J_{H\text{-}P\text{-}H}$ = 195.7 Hz);

IR, cm^{-1}: 2285 (P-H);

n_d^{20}=1.453; (lit.: 1.462 [13]);

b.p.: 34ºC (7 Torr); (lit.: 37-38ºC (9-10 Torr) [13]).

http://dx.doi.org/10.5772/53886

3. Result and discussion

3.1. Electrode reactions of elemental phosphorus

The electrochemical reduction of white phosphorus on a mercury dropping electrode in aprotic solvents is irreversible ($E_{1/2}$ = -1.55V in DMF) with the number of running electrons depending on the P_4 concentration. The formation of a radical-anion likely occurs at the very beginning of the electroreduction via one-electron transfer (Sch. 1):

$$P_4 + e^- \longrightarrow [P_4]^{\dot{-}}$$

Scheme 1. The electroreduction of white phosphorus.

Rupture of one or more P-P bonds followed by formation of P-C bonds, may then take place in the presence of proton donors via fast protonation of the electrogenerated anions. The electrochemical hydrogenation of elemental phosphorus and the synthesis of phosphine derivatives were studied in detail using various types of cathodes [14, 15]. The electrochemical reduction of red phosphorus in alkali solutions has received much less attention [16]. By these methods, PH_3 was synthesized in high yield using a water solution of NaOH (15-25%) and high temperature (70-100ºC). The yield of phosphine ranges between 60-83% at the lead cathode [16]. The application of a turbulent flow of the catholyte during the addition of a white phosphorus emulsion in a vertical electrolysis cell resulted in an improved efficiency in the electroproduction of PH_3 [17].

The electrochemical oxidation of white phosphorus coated on porous conductive matrices in neutral and acidic media was applied to produce stoichiometrically both phosphoric and phosphorous acids [18]. The indirect electrochemical formation of phosphorous oxyacids from P_4 took place in aqueous HX solutions (X = Cl, Br, I) [19, 20], but addition of non-aqueous solvents, such as benzene or chloroform, was sometimes necessary [19]. The oxidation of phosphorus by free halogen, previously liberated at the anode, is probably the first stage of the reaction to yield phosphorus halide derivatives, which are then hydrolysed to H_3PO_3.

The electrosynthesis of phosphorus and hypophosphorus acids was also reported from a suspension of red phosphorus in water in the presence of ethylenediamine and triethylamine [20]. The yields depended on the consumed electricity, although after ca. 9 F the content of phosphites and hypophosphites increased insignificantly. Nonetheless, the overall conversion to phosphorus species never exceeded 40-45 % with poor selectivity in either products.

3.2. Spin-adduct of the radical anion $P_4^{\cdot-}$

The problem of the selective cleavage of the P-P bonds in a white phosphorus molecule is very important in the chemistry of phosphorus-containing compounds [21, 22]. The radical character of the cathodic reduction of white phosphorus has been assumed earlier [23].

However, such products of activation of P_4 have not yet been detected. The purpose of the present study was to observe the primary products of cathodic reduction of a P_4 molecule. To create a necessary concentration of paramagnetic species in the resonator of an ESR spectrometer, we used a special cell with the helical working electrode [24]. The method of spin traps [25, 26] based on the reaction of electrochemically inactive α-phenyl-*N*-*tert*-butylnitrone (1), resulting in the stable nitroxyl radical (Sch. 2), was used to detect radical anion species. The stable nitroxyl radical can be identified by ESR spectroscopy.

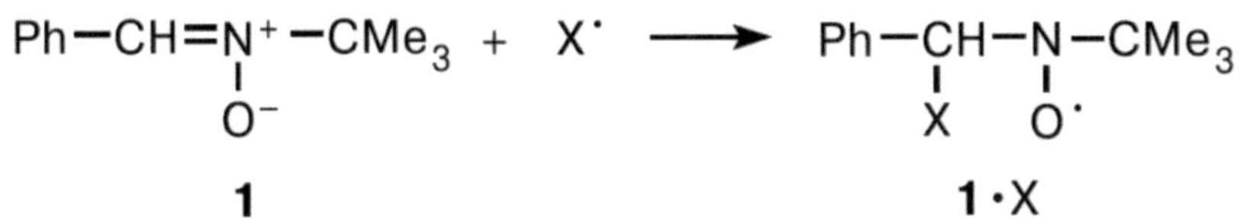

Scheme 2. Stable nitroxyl radical of α-phenyl-*N*-*tert*-butylnitrone.

The curves obtained by CVA at the Pt electrode for a solution of compound 1 in MeCN (0.01 M) and a solution of white phosphorus in a benzene-MeCN (1:1) mixture vs. 0.1 M solution of Et_4NBF_4. To exclude the formation of products of phosphorus oxidation and hydrolysis, all procedures were carried out in a thoroughly dehydrated solvent. It should be mentioned that the compounds containing the mobile hydrogen atom give no pronounced reduction peaks on Pt down to potentials of the supporting electrolyte discharge. The CV curves of a solution of white phosphorus at a potential of -1.5 V contain the irreversible peak corresponding to its reduction. Thus, the reduction peak close to the one-electron peak (estimated by the comparison with benzophenone used as standard) relates to the electron transfer to a white phosphorus molecule to form radical anions $P_4^{\bullet-}$.

There is a large data bank on the ESR spectra and magnetic resonance characteristics of spin-adducts of compounds belonging to various classes [27]. In particular, the spin-adduct of the short-lived phosphorus-centered radical $\bullet PHO_2^-$ and nitrone 1 was obtained [28] by the electrochemical oxidation of hypophosphite on the nickel electrode. The following magnetic resonance parameters of the spin-adduct were detected: $a_N = a_P = 15.91$ G, $a_{H\text{-}CH} = 1.99$ G, $a_{H\text{-}PH} = 3.21$ G, $g = 2.0060$. However, no spin-adducts of the $P4^{\bullet-}$ radical anion were studied.

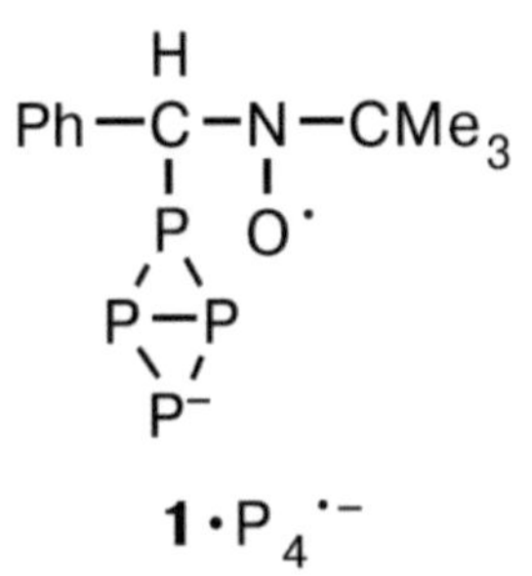

Scheme 3. Spin adduct **1**•P4˙⁻.

The reduction peak of nitrone 1 is by ~1 V more positive than the reduction peak of P_4 (-2.1V). Therefore, this trap was used for the detection of white phosphorus radical anions. The experimental ESR spectrum of the spin-adduct (Sch. 3) detected upon the reduction of white phosphorus (~16 mM) in the potentiostatic mode at a potential of -1.5 V and a 1 M solution of nitrone 1 in a benzene -MeCN (1:1) mixture at the Pt electrode vs. 0.1 M solution of Et_4NBF_4 is shown in Fig. 1.

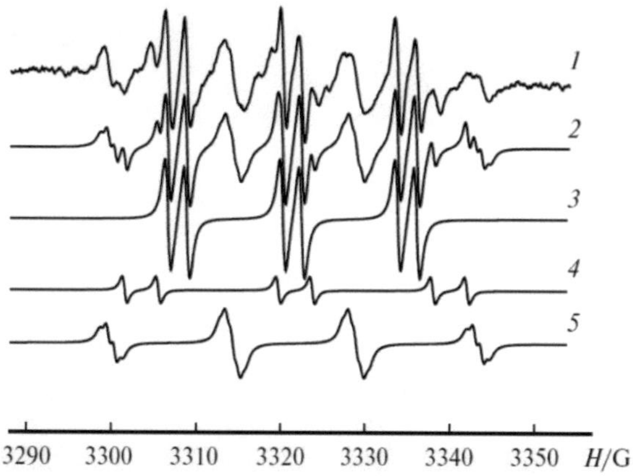

Figure 1. Experimental (*1*) and simulated overall (*2*) ESR spectra of the spin-adducts detected upon the reduction of white phosphorus and a 1 M solution of nitrone **1** in a benzene-MeCN(1:1) mixture vs. 0.1 M solution of Et_4NBF_4 after helium purging for 10 min and the ESR spectra of their constituent spin-adducts: **1 1·** (*3*), **1 1'·** (*4*), and **1** $P_4^{\cdot-}$ (*5*).

The ESR spectrum also exhibits two spin-adducts of free radicals of the spin trap 1 1 with the hyperfine coupling constants (HFC) $a_N = 13.59$ G, $a_H = 2.22$ G and $a_N = 18.60$ G, $a_H = 4.00$ G for the first and second adducts, respectively, with the ESR line widths $\delta H = 0.65$ G. The ESR spectrum of spin adduct 1•P4•– has the following parameters: $a_N = 14.10$ G, $a_{\beta\text{-P}} = 14.6$ G, $a_H = 0.80$ G, $a_{\gamma\text{-P}} = 0.78$ G, and $\delta H = 0.65$ G. The HFC constants $a_{\beta\text{-P}}$ and $a_{\gamma\text{-P}}$ were attributed to the phosphorus atom in the c-position and two equivalent phosphorus atom nuclei in the β-positions to the radical center, respectively. During several first minutes after the beginning of electrolysis, the lines of the experimental spectrum begin to broaden and the pattern changes gradually, indicating the formation of polymer products. Thus, the radical anion $P_4^{\cdot-}$ has been fixed and identified for the first time as spin-adduct with a α-phenyl-N-*tert*-butyl nitrone during electrochemical reduction of P_4.

3.3. The formation of polymer products upon the electrochemical reduction of white phosphorus

In the following conditions: DMF vs. 0.1 M solution of Et_4NBF_4 at E = -2.1 V and $C_{P4} = 50$ mM is confirmed by the microscopic images of the electrode surface (Figs. 2 and 3). The view of the polyphosphorus films depends on the electrolysis conditions and duration. The polyphosphorus products formed at the Pt electrode in DMF are rapidly dissolved and desorbed to the electrolyte volume. At the same time, on the lead cathode in DMF they form a black sponge nanoporous film (see Figs. 2 and 3). In a water containing electrolyte, a black polyphosphorus conducting film is also formed, but the structure of polyphosphides is filamentous and needle-like in this case (see Fig. 3). The thickness of the polymer filaments is ~1 μm.

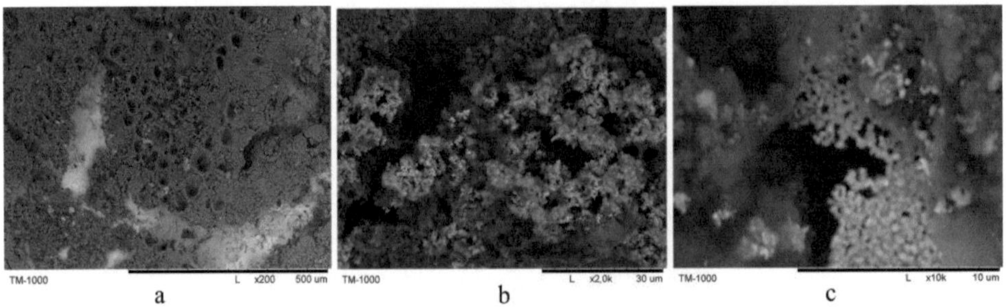

Figure 2. View of the polyphosphorus films upon the electrochemical reduction of P4 in DMF vs. 0.1 M solution of Et_4NBF_4 detected with a HITACHI TM-1000 microscope with a magnify cation of 200 (a), 2000 (b), and 10000 times (c).

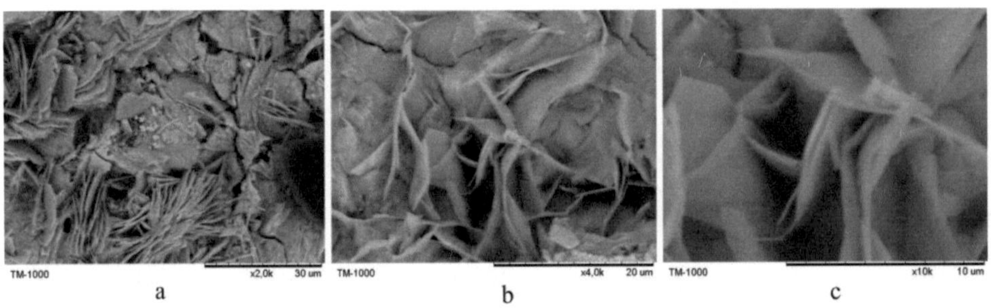

Figure 3. View of the polyphosphorus films upon the electrochemical reduction of P4 in an aqueous solution of HCl detected with a HITACHI TM-1000 microscope with a magnification of 2000 (a), 4000 (b), and 10000 times (c).

3.4. CVA of white phosphorus

The electrochemical reduction of white phosphorus was not studied in details up to now, the mechanism of its activation and transformation with P-P bond cleavage under electron transfer on electrodes of the various nature was not clear. The data on potentials of P_4 reduction on glassy carbon and mercury were published earlier [23], the ways of transformation of white phosphorus under the complexes of nickel, including generated electrochemically were described [29-31].

Research of an opportunity and a mechanism of heterogeneous activation of white phosphorus on various electrodes, finding-out of the factors determining ways of transformation of white phosphorus in polyphosphoric cycles, detection of short-living intermediates with the purpose of management of capture processes of highly reactive phosphoric oligomer forms $[P_n]^{m-}$ by various substrata for example, olefins, at their joint reduction, represent a special interest. Simultaneous voltammetry (with linear or cyclic potential scan) and ESR allow to throw light on details of the mechanism of electron transfer processes.

The purpose of the present work is the establishment of laws of reduction of white phosphorus on various electrodes (platinum (Pt), glassy carbon (GC), lead (Pb)), the mechanism of electron transfer and the nature of intermediates with application of methods of simultaneous voltammetry and ESR-spectroscopy.

It was revealed that potentials of P_4 reduction and also currents and accordingly number of transferable electrons differ a little on various electrodes (Figs. 4-6 and Table 1). So, on glassy carbon electrode (Fig. 4) the reduction wave of white phosphorus is close to one-electronic, and on platinum (Fig. 5) and lead ones (Fig. 6) it is a little bit more one-electronic at use of benzophenon as the standard. The current of reduction is directly proportional to concentration of a substratum. Potentials of reduction settle down in the following sequence-the least negative is on a lead electrode, then on glassy carbon electrode, and the most negative-on platinum one.

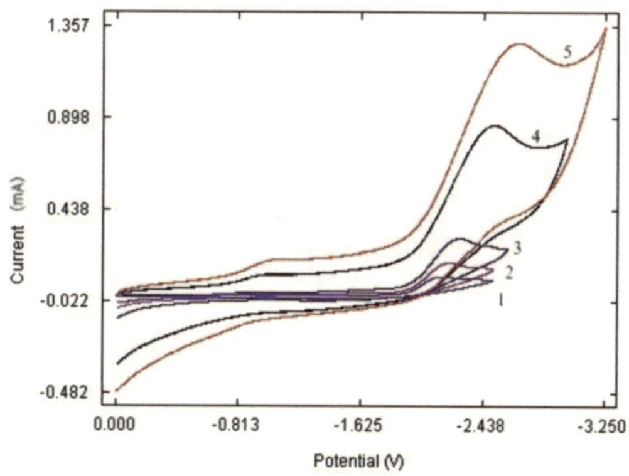

Figure 4. CVAs of white phosphorus solutions (10 mM) in DMF on GC electrode. Potential: V vs. SCE. Scan rate (bottom-up): 1 – 50 mV/s, 2 – 100 mV/s, 3 – 300 mV/s, 4 – 1000 mV/s, 5 – 10000 mV/s, 6 – 25000 mV/s.

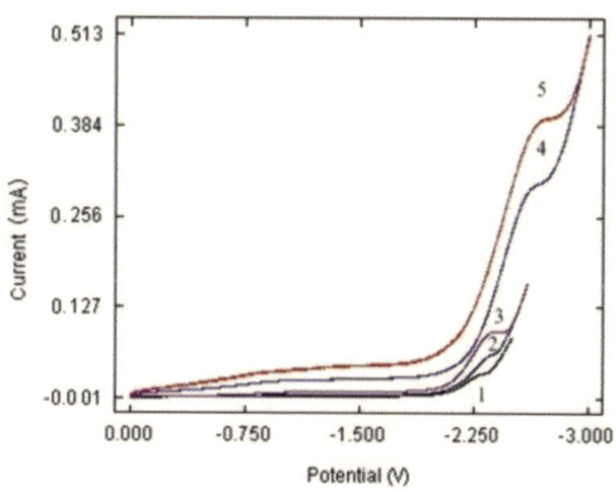

Figure 5. CVAs of white phosphorus solutions (10 mM) in DMF on Pt electrode. Potential: V vs. SCE. Scan rate (bottom-up): 1 – 100 mV/s, 2 – 300 mV/s, 3 – 1000 mV/s, 4 – 10000 mV/s, 5 – 25000 mV/s.

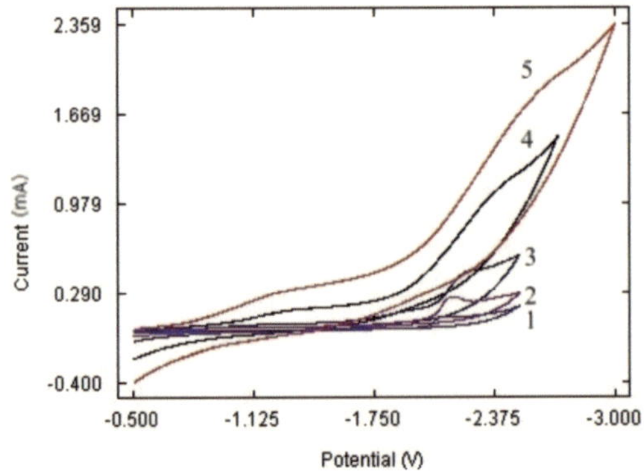

Figure 6. CVAs of white phosphorus solutions (10 mM) in DMF on Pb electrode. Potential: V vs. SCE. Scan rate (bottom-up): 1 – 50 mV/s, 2 – 100 mV/s, 3 – 300 mV/s, 4 – 1000 mV/s, 5 – 10000 mV/s, 6 – 25000 mV/s.

Electrode	n_e	-E, V
GC	1.01	2.16
Pt	1.30	2.30
Pb	1.27	2.12

Table 1. Characteristics of reduction pics of white phosphorus. DMF, 10 mM P_4, 0.1 M Et_4NBF_4, Scan rate: 100 mV/s. Potential: V vs. SCE, standard-benzophenone.

An anodic component of wave is observed on the glassy carbon electrode at high potential scan at return scanning that can testify to some stability of a product of primary carry of an electron on a molecule of white phosphorus $[P_4]^{\overline{\bullet}}$. The process is completely irreversible on other electrodes. However it is visible by detailed consideration of character of voltammetric curves, that the peak has the complex form on GC at record of several cycles (Figs. 7a and 7b). Last is shown more distinctly at record of curves in coordinates «semi derivative of a current-potential» (Fig. 7b). The form of peak can be connected to polymerization of phosphorus on the electrode.

Coefficient of diffusion of white phosphorus and coefficient of transfer have been estimated (Table 2) on the basis of received voltammetric curves. Calculation was carried out on Delahey equation for irreversible processes [32]:

$$i_p = \left(2,99\cdot10^5\right)\cdot n\cdot\left(\alpha n_a\right)^{1/2}\cdot S\cdot D^{1/2}\cdot v^{1/2}\cdot C_o \qquad (1)$$

where

i_p - current, A;

n - number of electrons;

α - electron transfer coefficient;

S - electrode surface, cm^2;

D - coefficient of diffusion, cm^2/s;

v - scan rate, V/s;

C_o - concentration, mol/cm^3.

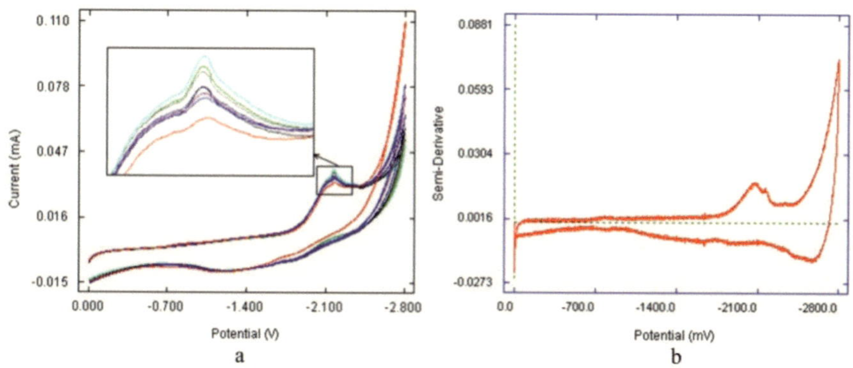

Figure 7. CVAs of white phosphorus solutions (5 mM) in DMF on GC electrode. Potential: V vs. SCE. Scan rate: 100 mV/s. a – CVA from 0 to -2.8 V, b – dependence of semi-derivative of a current from potential.

Linear dependences of white phosphorus reduction current from $\sqrt{v}$ (v – scan rate) on all electrodes are observed. That process is diffusion controllable (Fig. 8).

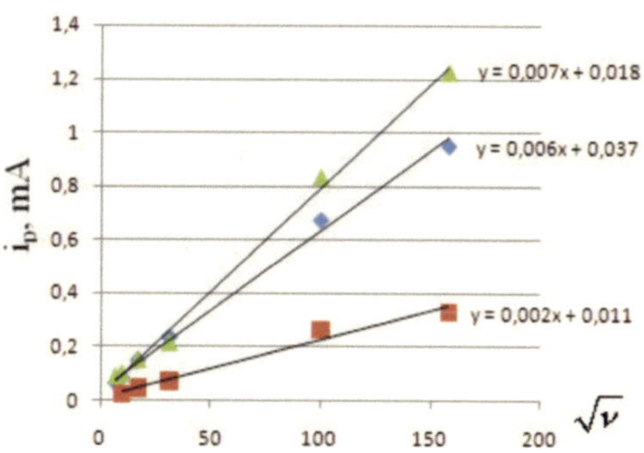

Figure 8. White phosphorus reduction peak current dependences on $\sqrt{v}$ (v – scan rate) on various electrodes: GC, Pt, and Pb.

The values of the transfer coefficient according Eq. (1):

$$\alpha = \frac{RT}{F} \cdot \frac{1{,}85}{E_{p/2} - E_p} \qquad (2)$$

$$D = \left(\frac{i_p}{\left(2{,}99 \cdot 10^5\right) n \cdot \sqrt{\alpha n_a} \cdot \sqrt{\nu} \cdot C_O} \right)^2 \qquad (3)$$

Electrode	C_o, mM	Scan rate, mV/s	ΔE_p, mV	α	D, cm²/s
GC	5	100	-156	0.304	1.82×10^{-5}
Pb	5	100	-52	0.879	6.01×10^{-6}
GC	10	100	-123	0.386	2.06×10^{-5}
Pb	10	100	-43	0.931	1.39×10^{-6}
Pt	10	100	-112	0.424	2.67×10^{-5}

Table 2. Calculation of coefficient of diffusion on various electrodes.

The average value of the coefficient of diffusion of white phosphorus calculated in DMF is 2 × 10^{-5} cm^2/s.

3.5. Electrosynthesis of primary phosphines from alkenes and white phosphorus

Primary phosphines are valuable starting materials in many chemical reactions. Because of their importance as precursors or active intermediates in many reactions, there is a growing interest in a new preparation method from an available raw material such as elemental phosphorus. Replacement of the present-day processes for preparation of basic organophosphorus compounds now based on reactions of phosphorus chlorides is becoming increasingly important. The existing processes are environmentally dangerous, energy-demanding and cause problematic waste streams. The creation of chlorine- and waste-free processes aimed at the obtaining of some starting compounds, namely, phosphites, phosphates, amides and tertiary phosphines, etc., based on elemental phosphorus is an alternative of the organophosphorus compounds synthesis.

The syntheses of primary phosphines usually involve expensive, multi-step and long-term procedures, such as: the use of reduction of phosphorus dihalides with $LiAlH_4$, preparation from metal phosphides and alkyl or aryl halides, hydrolysis of alkyl and arylphosphorus dihalides, reaction under Friedel-Crafts conditions, pyrolysis of biphosphines and triphosphines, etc. [10,33,34]. Selective method of primary phosphine preparation directly from white phosphorus escaping, traditional stages of its chlorination, with formation of unique phosphorus compounds with phosphorus-carbon bond is not known up to now.

Reactions of olefines with phosphorylating agents are not selective as a rule. It is known [11], that styrene reacts with PH_3 at 70ºC and a pressure of 28-30 atmospheres under the action of radical initiators. This reaction results in the formation of the mixture of primary, secondary, and tertiary phosphines with the yields of 6-36%. The way of primary phosphines preparation from styrene or α-methylstyrene and PH_3 in superbasic medium (KOH/DMSO) with the yield of 20-30% are also well known [9].

Electrochemical methods of action on white phosphorus are found to be very promising in many cases, since they allowed elaborating the procedures of selective synthesis of phosphorus acid esters, triphenylphosphines and other products [35-37].

However up to now one failed to elaborate an approach to synthesize from P_4 phosphine derivatives with one or two P-H bonds, e.g. primary or secondary phosphines R_2PH, RPH_2, tertiary phosphines other than triphenylphosphine, phosphorus acids H_3PO_3, H_3PO_2 and others, being important precursors in phosphorus chemistry. The main problem of all known reactions concerns either low yield of a product due to the formation of nonreactive polyphosphides and consequently low phosphorus conversion, or the use of expensive reagents, such as rhodium complexes [38-40]. Our attention was attracted by rather old publications on phosphine electrolytic production, which were carried out already in the sixties 20th centuries and for some reason did not receive further progression. Thus, the cathodic reduction of white phosphorus in aqueous solutions on metals with high hydrogen overvoltage was shown to result in the formation of PH_3 with yield up to 95%; these results were patented in Germany, USA, and Great Britain [41-47]. Recently, some technological refinement for this process was suggested by Japanese researchers, who patented a turbulent generating process of PH_3 [48]. In the sixties 20th centuries and at the beginning of the seventies 20th centuries several attempts were undertaken to use electrochemically generated PH_3 for subsequent synthesis on the base of its reactions in situ, however, the obtained results were not encouraging as they resulted in complex mixtures of products (e.g. with styrene) [14-16] and/or low yields.

It is worth noting that the direction of white phosphorus conversion into phosphine is sufficiently well worked out by patents, and the present task consists in creation of such conditions, which would allow instant (as formed) conversion of PH_3 and other phosphine intermediates into the derivatives with P-H bonds, e.g. organic phosphines. This will allow avoiding the accumulation of intermediate toxic and dangerously explosive phosphine, converting it into undetectable "conventional intermediate":

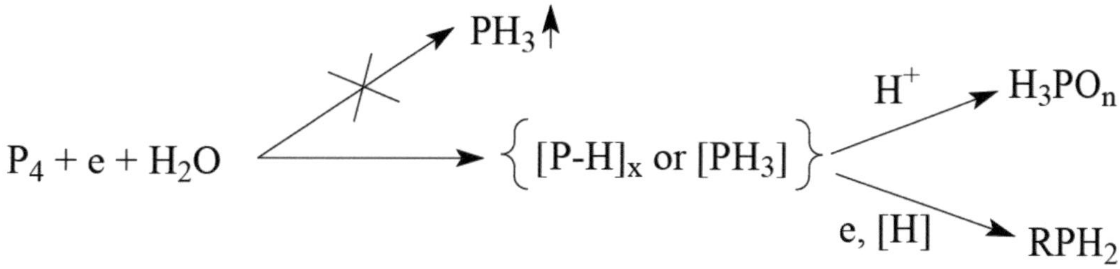

Scheme 4. Electrolysis P_4 in aqueous solution.

Joint electrolysis of white phosphorus emulsion and alkene in the aqueous acetic buffer solution results in the formation of just primary phosphine in these conditions (Table 3):

$$1/4P_4 + CH_2{=}CH\text{-}R \xrightarrow{3H^+,\ 3e} H_2P\text{-}CH_2CH_2\text{-}R$$

Scheme 5. Formation primary phosphine from P_4 in aqueosu acetic buffer solution.

The mechanism of phosphine with P-C bond formation is not quite clear. It should be assumed as follows, that phosphorus centered radicals and radical-anions are generated from white phosphorus under conditions of electrochemical reduction. Radicals and radical-anions can add to weakly electrophilic alkenes (e.g. styrene) by the nucleophilic mechanism, and to nucleophilic alkenes (alkyl ethene, phenyl ethers)-by the radical mechanism.

Electrochemical reduction of P_4 molecule in protogenic conditions is known to take place at the cathode with high hydrogen overvoltage, for example, at the lead cathode [35-37]. The electrochemical rupture of the P-P bonds, resulting in the formation of phosphine, is provided by the presence of active proton donors through the protonation of intermediates:

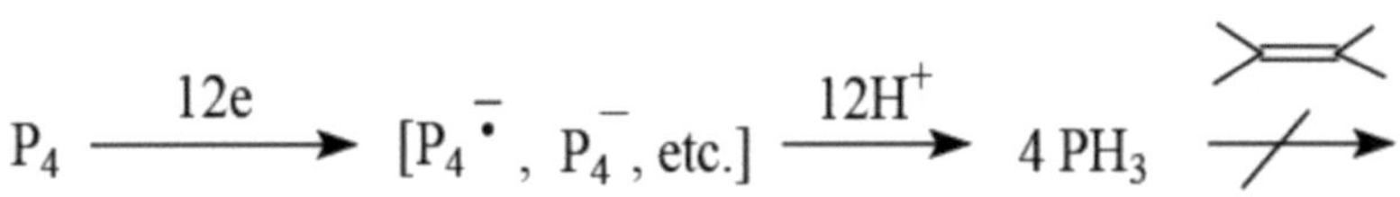

Scheme 6. Electrochemical reduction of P_4 molecule.

But on the other hand, phosphine does not react with alkene without catalyst. Phosphine can be added to C=C bonds both by ionic and radical mechanisms in the presence of initiators. Alkenes of different structure react with phosphine under rather rigid conditions (60-90ºC, 30-47 atm, acid catalyst; or in the superbasic medium such as KOH/DMF) [34]. Apparently, under electrolysis conditions the reaction under investigation is initiated by phosphine reduction to $^{-}PH_2$, occurring at the electrode, or proceeds through the intermediate phosphides or radical and radical-anions formed at white phosphorus reduction. Generally, it is impossible to exclude proceeding of several competitive reactions of phosphide-anion formation, both its protonation in solution and reduction of P-H bond at the electrode, and target addition of phosphide-anion to alkene.

Voltammetric (coulometric) studies of white phosphorus reduction were performed at mercury electrode in alcohol medium [49] and have shown the process to proceed at substantial negative potentials, the calculation of a number of electrons through the peak height giving the value of n close to unity in the concentration range from 0.05 to 1.6 mM, what practically coincides with white phosphorus solubility limit in these media. At glass carbonic electrode in aprotic media (DMFA, CH_3CN) with modest addition of benzene the process proceeds at -2.20 V ($Ag/AgNO_3$), at concentration 5 mM a number of electrons being even slightly smaller than unity-0.7 with benzophenone standard [30]. At lead electrode white phosphorus reduces at similar values of potential (-2.20 V vs. $Ag/AgNO_3$; -1.86 V vs. SCE) and current. To understand the mechanism of white phosphorus reduction one may consider the variation

http://dx.doi.org/10.5772/53886

of phosphorus concentration with the amount of transmitted current according to the A.P. Tomilov procedure [14,50]. In doing so the methanol solution admixed with benzene and styrene, which is not indifferent to phosphorus in conditions of reduction, is changed for inert solvent, e.g. dimethylformamide admixed with benzene or toluene to improve solubility. According to the calculation, the loss of phosphorus at the beginning of electrolysis corresponds to the one-electron process, and then phosphorus concentration becomes so low, that the voltammetry does not provide reliable results. Visual observation demonstrates after nearly 0.5 F/R the formation of dark oily liquid not solidifying (hardening) at room temperature, insoluble in aqueous media and instantly flaring up in the air; this liquid is to a large measure adsorbed at a leaden cathode and disappears after transmitting of 3 F/R. In aprotic media a considerable amount of white phosphorus converts into insoluble yellow polyphosphides frequently observed at P_4 exposure to nucleophilic reagents. In that way the loss of white phosphorus in a solution occurs much faster than it is required for three-electron process (Fig. 9).

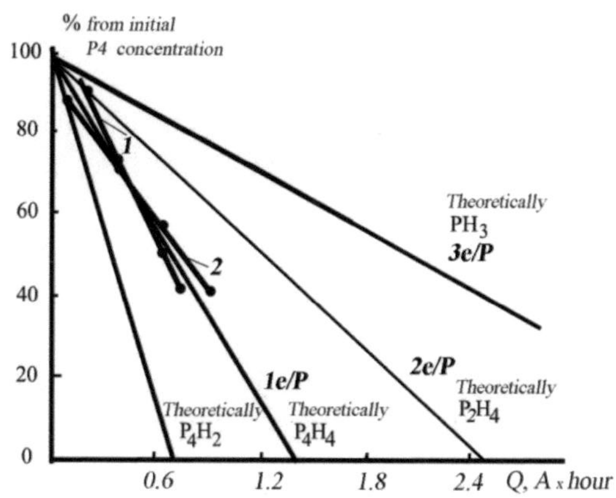

Figure 9. Measurement of phosphorus concentration in KAc/HAc solution in DMF with a mixture of benzene (1) and toluene (2) in the process of cathodic reduction at lead electrode.

Thus, one may conclude that first two of three possible reactions do not occur at the initial stage of electrolysis:

$$P_4 + 12e \xrightarrow{+12H^+} 4PH_3 \uparrow \qquad 3e/P$$

$$P_4 + 8e \xrightarrow{+8H^+} 2P_2H_4 \qquad 2e/P$$

$$P_4 + 4e \xrightarrow{+4H^+} P_4H_4 \qquad 1e/P$$

Scheme 7. Stage of electrolysis of P_4.

Phosphorus one-electron reduction results in the formation of phosphorus hydride P_4H_4. Cyclic P_4H_4 is not available in literature, though its existence is supposed in the series of hydrides of the $(PH)_x$ type [10,51]. P_4H_4 is insoluble in water; it is known to have melting temperature 99ºC, boiling temperature 56ºC [51]. Evidently, at the early stages of electrolysis there occurs the formation of hydrides, which are water-insoluble, but are well soluble in phosphorus, with the curing temperature being decreased and coloring being darkened. Liquid phosphorus hydride, withdrawn from the electrolyzer, decomposes at light under water with solid yellow polyphosphoric products being formed. It is interesting, that in the considered aqueous solutions white phosphorus is practically insoluble, and however, nevertheless it efficiently reduces at leaden cathode at P_4 emulsification. This phenomenon is explained by the so-called "wick effect": molten phosphorus lifts vertically along the cathode similarly to the liquid along the wick, moistening its whole surface [51].

The formation of secondary or tertiary phosphines was not observed even at significant alkene excess in initial mixture. According to NMR ^{31}P spectrum, the only by-products discovered in the reaction mixture in aqueous part of electrolyte were inorganic acids of phosphorus. The proposed method is characterized by the following advantages: mild conditions (room temperature) of the process and the one-step way of primary phosphine preparation directly from white phosphorus escaping all traditional stages of its functioning.

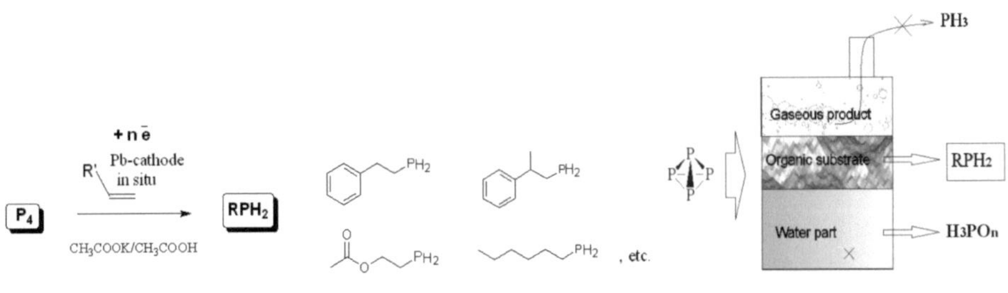

Scheme 8. Products of electrolysis of white phosphorus in water-phosphines and its derivatives (organic and inorganic).

It should be noted that the use of some alkenes does not afford an opportunity of the desired phosphines formation (Table 3).

The results, obtained in the course of electrosynthesis, are explained on the basis of the voltammetric data for substrates. According to cyclic voltammetry data, reduction potentials of alkene, successfully reacting with P_4 in joint electrolysis (first four lines in Table 3), are more negative, than the potential of white phosphorus reduction (E = -2.2V). Therefore, the process proceeds through P_4 reduction to the desired products. Other alkenes (last four lines in Table 3) are reduced more easily than phosphorus, what hinders the primary phosphines formation. Limitation of the process is caused by obtaining only phosphines with H_2P-$(CH_2)_2$-bonds.

http://dx.doi.org/10.5772/53886

Substrat	Reduction potential, E_{red}, V		Product	Yield (%)
	GC cathode	Lead cathode		
Styrene 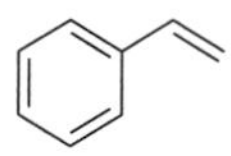	No reduction	-2.95	2-phenylethylphosphine 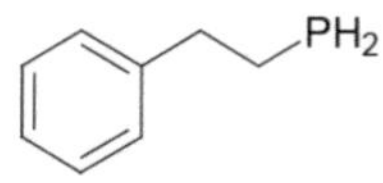	31
α-methylstyrene 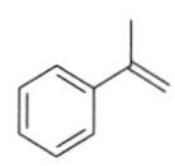	No reduction	-3.05	2-phenylpropylphosphine 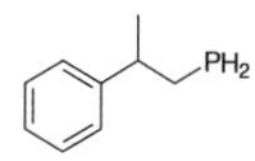	46
vinylacetate 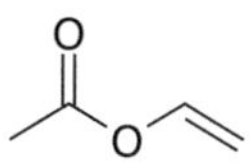	No reduction	-2.90	2-acetoxyethylphosphine 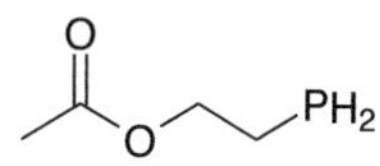	22
hexane-1 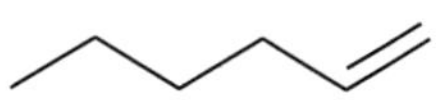	No reduction	No reduction	n-hexylphosphine 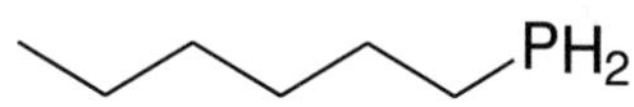	23
2-vinylpyridine	-2.10	-	-	0
acrylamide	-2.05	-	-	0
phenylacetylene	-2.17	-	-	0
vinylnaphtalene	-1.96	-	-	0

Table 3. Reduction potentials of initial alkenes and yields of their phosphorylation products.

3.6. Electrocatalyzed functionalization of white phosphorus in the presence of nickel complexes

The electrocatalytic functionalization of white phosphorus has been also accomplished using transition metal complexes as catalysts [52-55].The activation of P_4 takes place under mild conditions promoted by the electrogeneration of a carbanion in the presence of P_4 [30,53-55]. The reaction has been documented for Ni(0) species stabilised by 2.2′-bpyridine (bpy) electrogenerated from Ni(II) complexes. The reduced Ni(0) species react with organic halides to give σ-organonickel complexes which may further participate in several organic elaborations of different substrates, including phosphorus [22,35,53-55].In such a case, the electrolysis is carried out using a soluble electrode (Al, Mg, Zn) in an undivided cell with a phosphorus emulsion in DMF or acetonitrile containing an organic halide and a Ni(II) complex, $[Ni(bpy)_3](BF_4)_2$, as catalyst. Under these experimental conditions, white phosphorus may be efficiently converted to phosphines and phosphine oxides [53-55].

Mechanistic studies were carried in the specific case of electrocatalytic arylation of white phosphorus [53-55]. From these investigations, it was determined that the Ni(0) complex, initially obtained via reduction of Ni(II), oxidatively adds the organic halide to give [NiX(Ar)(bpy)] species, which mediate the catalytic formation of P-C bonds.

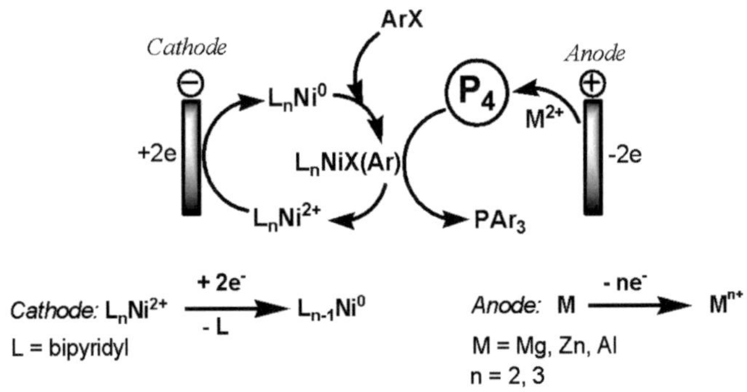

Scheme 9. Representation of the nickel electrocatalyzed arylation of white phosphorus.

The nature of the soluble anode drastically influences the final product although such a behaviour denies any simple interpretation (Sch. 10). Then, the use of a zinc anode leads to complete conversion of P_4 to soluble OPCs, mainly, tertiary phosphines, while an aluminium anode electrogenerates the phosphine oxide. Cyclic polyphosphorus compounds, such as $(PhP)_5$, are produced when a magnesium anode is used [35,53-55].

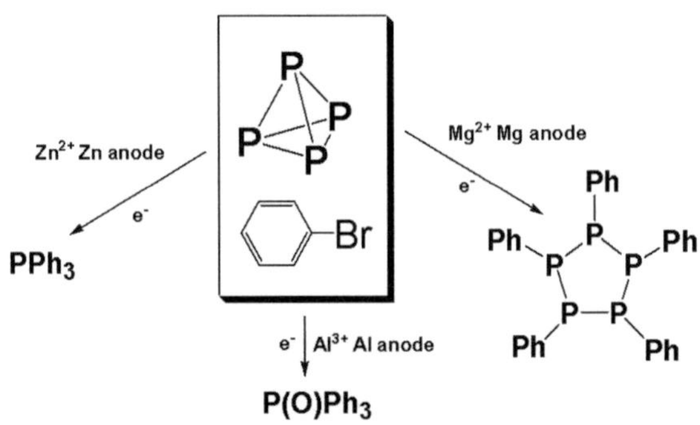

Scheme 10. Electrode dependence of the product resulting from the electrocatalyzed arylation of white phosphorus with bromobenzene.

These processes are intriguing because they underpin the high potentiality of metal electrocatalysis in bringing about the alkylation and arylation of white phosphorus under mild conditions. Remarkably, these processes combine a high efficiency in OPC formation with a total control of the product selectivity depending on the careful choice of the metal anode.

We have developed essentially new approach to triphenyl- or perfluoroalkylphosphine preparation from white phosphorus, using not divided electrolyzer and activation of white phosphorus under the action of zinc compounds:

http://dx.doi.org/10.5772/53886

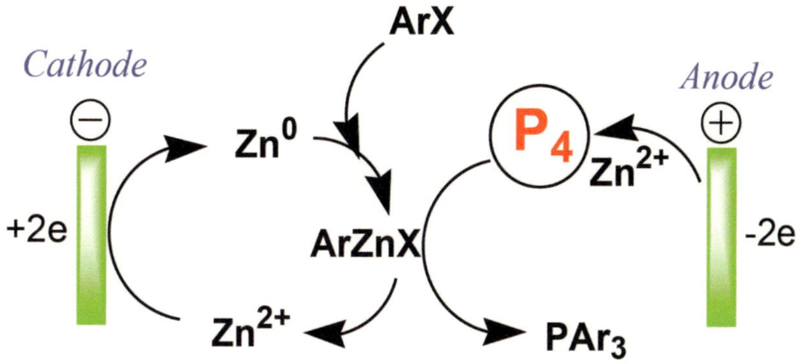

Scheme 11. Representation of the Zn catalyzed arylation of white phosphorus.

Cyclic regeneration of Zn-catalyst takes place at the cathode, reactive sigma-complex forms in the bulk of the solution. The σ-complex attacks the P_4 molecule resulting in tertiary phosphines as target product.

4. Conclusions

Combining electrochemistry with homogeneous catalysis offers many interesting opportunities for the direct application to the synthesis of organophosphorus derivatives. As the electrochemical equipment and the adopted procedures become more and more simple, it is expected that electrochemical methods will successfully compete with the conventional synthetic methods. The former allow carrying out the wished synthesis avoiding the use of chlorine and under mild conditions with high reaction rates and good productivities. They also permit a fine-tunable control of the process resulting therefore in high or sometimes complete selectivity. Thus, electrosynthesis allows single-stage converting of white phosphorus and alkenes under mild conditions into primary products, being the single products with phosphorus-carbon bond. Inorganic hypophosphorus acid is found to be a by-product. The choice of alkenes is defined by their smaller electrochemical activity at glass-carbon electrode they are not reduced in the accessible region, while at leaden electrode they reduce at lower negative potentials.

The $P4^{\bullet-}$ radical anion was detected for the first time in the potentiostatic mode by the ESR method as the spin-adduct with α-phenyl-*N-tert*-butylnitrone in the electrochemical reduction of white phosphorus in the electrolysis cell with the helical platinum working electrode. The present study confirms the radical character of the intermediates of white phosphorus.

The base to realization of the developed processes of organophosphorus compounds preparation from elemental phosphorus at the technological level was incorporated as a result. Scientific bases highly effective, resource saving and ecologically safe technology of electrosynthesis of the major classes of OPC were created. The block diagram of synthesis and separation of a target product was made. The technological instruction of process was developed.

There are many interesting promising synthetic applications combining electrochemistry with homogeneous catalysis.

The interest to the organophosphorus compounds electrosynthesis on the basis of white phosphorus is also caused by a number of advantages compared to common chemical methods:

- No additional chemicals; a possibility to carry out the reaction in practically closed system with minimum amount of cyclically regenerated reagents;
- Greater product selectivity and yield
- Reduced or no disposal cost
- Recovery, recycling of astes/pollutants
- Low capital costs/low operating costs.

Essentially, new approach to OPC synthesis, having practical value, from white phosphorus in electrocatalytic conditions, first of all with P-C or P-H bonds based on effective universal technology of electrosynthesis of lines of OPC compounds were created.

The electrochemistry is a powerful synthetic method for preparing a wide range of phosphorus compounds.

Author details

Yu. G. Budnikova and S. A. Krasnov

A. E. Arbuzov Institute of Organic and Physical Chemistry Kazan Scientific Center, Russian Academy of Sciences, Russia

References

[1] Kolbe H. Untersuchungen über die Elektrolyse organischer Verbindungen. Annalen der Chemie und Pharmacie. 1849;69(3) 257–372.

[2] Degner D. Organic electrosyntheses in industry. Topics in Current Chemistry; 1988,148.

[3] Budnikova Yu.H. Electrosynthesis of organic compounds. Ecologically safe processes and design of new synthetic methods. Rossijskij Khimicheskij Zhurnal (Zhurnal Rossijskogo Khimicheskogo Obshchestva Im. D.I. Mendeleeva). 2005;49(5) 81-93.

[4] Budnikova Yu.H. Metal complex catalysis in organic electrosynthesis. Russian Chemical Reviews. 2002;71(2) 111-139.

[5] Utley J. Trends in organic electrosynthesis. Chemical Society Reviews. 1997;26(3) 157-167.

[6] Kargin Yu. M., Budnikova Yu. H. Electrochemistry of organophosphorus compounds. Russian Journal of General Chemistry. 2001; 71(9) 1393-1421.

[7] Amatore C., Jutand A. Mechanistic and kinetic studies of palladium catalytic systems. Journal of Organometallic Chemistry. 1999;576 (1-2) 254-278.

[8] Kadirov M. K., Odivanov V. L., Budnikova Yu. G. Pribory I tekhnika eksperimenta [Experimental Instruments and Technique], 2007;1(151) (in Russian).

[9] Gusarova N.K., et al. Synthesis primary phosphines from phosphine and arylethylenes. Russian Chemical Bulletin. 1995; 8 1597-1598.

[10] Corbridge D.E.C. Phosphorus 2000. Chemistry, Biochemistry&Technology. Elsevier, Amsterdam/Lausanne/New York/Oxford/Shannon/Singapore/Tokyo; 2000.

[11] Rauhut M.M., et al. The free radical addition of phosphines to unsaturated compounds. Journal of Organic Chemistry. 1961;26(12) 5138-5145.

[12] Pass F., Schindlbauer H. Organische Verbindungen des Phosphors. I. Mitt. - Über die Darstellung primärer Phosphine durch reduktive Methoden. Monatshefte für Chemie. 1959;90(2) 148-156.

[13] Knunjanz I.L. and Sterlin R.N., Doklady AN SSSR (Russ.).1947;1 47-50.

[14] Shandrinov N.Ya., Tomilov A.P. Electrocemically reduction of phosphorus on a lead electrode. Electrochemistry. 1968;4 237.

[15] Osadchenko I.M., Tomilov A.P. Electrochemical synthesis of hydrogen phosphide. Zhurn. prikl. Khimii. 1970;43 1255

[16] Osadchenko I.M., Tomilov A.P. Syntesis lower acids of phosphorus by electrolysis of suspension red phosphorus. Electrochemistry. 1993;29(3) 406.

[17] Tsuchiya H., Otsuji A., Sakagami Yu., Japan Patent 01139781 A2 19890601 Heisei, 1989.

[18] Barry M.L., Tobias Ch. W., Electrochimical Technology. 1966;4 502.

[19] US Patent 4,021,321 (Cl.204-103; C25B1/22), (1977).

[20] US Patent 4,021,322 (Cl.204-103; C25B1/22), (1977).

[21] Peruzzini M., Abdreimova R. R., Budnikova Y., Romerosa A., Scherer O. J., Sitzman H. Functionalization of white phosphorus in the coordination sphere of transition metal complexes. Journal of Organometallic Chemistry. 2004;689 4319-4331.

[22] Milyukov V.A, Budnikova Yu.H., Sinyashin O.G. Organic chemistry of elemental phosphorus. Russian Chemical Reviews. 2005;74 781.

[23] Budnikova Yu. G., Krasnov S. A., Sinyashin O. G. Design of ecologically safe and science intensive electrochemical processes. Russian Journal of Electrochemistry. 2007;43(11) 1223-1228.

[24] M. K. Kadirov, Patent. RF 69 252. Byul. Izobret. [Invention Bulletin]. 2007; 34 (in Russian).

[25] Janzen E. G. Spin trapping. Accounts of Chemical Research. 1971;4(1) 31-40.

[26] Bard A.J., Gilbert J.C., Goodin R.D. Application of spin trapping to the detection of radical intermediates in electrochemical transformations [30]. Journal of the American Chemical Society. 1974;96(2) 620-621.

[27] The National Institute of Environmental Health Sciences. Search the Spin Trap Database. http://epr.niehs.nih.gov.

[28] Zeng Y., Zheng Y., Yu Sh., Chen K., Zhou Sh. An ESR study of the electrocatalytic oxidation of hypophosphite on a nickel electrode. Electrochemistry Communications. 2002;4(4) 293-295.

[29] Budnikova Yu.H., Tazeev D.I., Trofimov B.A., Sinyashin O.G. Electrosynthesis of nickel phosphides on the basis of white phosphorus. Electrochemistry Communications. 2004;6(7) 700-702.

[30] Budnikova Yu.G., Tazeev D.I., Kafiyatullina A.G., Yakhvarov D.G., Morozov V.I., Gusarova N.K., Trofimov B.A., Sinyashin O.G. Activation of white phosphorus in the coordination sphere of nickel complexes with σ-donor ligands. Russian Chemical Bulletin. 2005;54(4) 942-947.

[31] Budnikova Yu.G., Tazeev D.I., Gryaznova T.V., Sinyashin O.G. Novel high-efficiency ecologically safe electrocatalytic techniques for preparing organophosphorus compounds. Russian Journal of Electrochemistry. 2006;42(10) 1127-1133.

[32] Bard A.J., Faulkner L.R., Electrochemical Methods: Fundamentals and Applications, second ed., New York, 2001.

[33] Kosolapoff G.M. and Maier L. Organic Phosphorus Compounds. John Wiley and Sons, New York/London/Sydney/Toronto. 1976;7 871 pp.

[34] Trofimov B.A., et al. Phosphine in the synthesis of organophosphorus compounds. Russian Chemical Reviews. 1999;68(3) 215-227.

[35] Budnikova Yu. H., Yakhvarov D., Sinyashin O.G. Electrocatalytic eco-efficient functionalization of white phosphorus. Journal of Organometallic Chemistry. 2005;690(10) 2416-2425.

[36] Budnikova Yu. H., Kargin Yu.M., Romakhin A.S., Sinyashin O.G. Patent Russian Federation № 2199545.2003.

[37] Budnikova Yu. H., Yakhvarov D.G., Sinyashin O.G. Patent Russian Federation № 2221805. 2004.

[38] Di Vaira M., Frediani P., Seniori Costantini S., Peruzzini M., Stoppioni P., Dalton Trans. Easy hydrolysis of white phosphorus coordinated to ruthenium. Dalton Transactions.2005;(13) 2234-2236.

[39] Barbaro P., Peruzzini M., Ramirez J. A., Vizza F. Organometallics 18 (1998) 2376.

[40] Barbaro P., Peruzzini M., Ienco A., Mealli C., Scherer O. J., Schmitt G., Vizza F., Wolmershäuser G. Activation and Functionalization of White Phosphorus at Rhodium: Experimental and Computational Analysis of the [(triphos)Rh (η 1:η2-P4RR′)]Y Complexes (triphos = MeC(CH2PPh2)3; R = H, Alkyl, Aryl; R′ = 2 Electrons, H, Me). Chemistry - A European Journal. 2003;9(21) 5195-5210.

[41] Patent USA 3109790 (1963); C.A. 60, 2552. 1964.

[42] Patent USA 3109791 (1963); C.A. 60, 2552. 1964.

[43] Patent USA 3109792 (1963); C.A. 60, 2552. 1964.

[44] Patent. Germany 1210424 (1966); C.A. 64, 13767d. 1966.

[45] GB Patent 1042391. 1966.

[46] Patent Germany 1210425 (1966); C.A.64, 13767f. 1966.

[47] Patent Germany 1210426 (1966); C.A.64, 12198g. 1966.

[48] Patent Japan 01139781 A2 19890601 Heisei. 1989.

[49] Tomilov A.P., Mairanovskii S.G., Fioshin M.Ya., Smirnov V.A. Electrochemistry of Organic Compounds. «Khimiya. Leningrad», 1968 591.

[50] Tomilov A.P., Osadchenko I.M. Journal AnalitycalChemistry (Russian). 1966; 1498.

[51] Osadchenko I. M., Tomilov A.P. Electrocemically synthesis of phosphine oxide. Russian Chemical Review. 1969;38(6) 1089-1107.

[52] Budnikova Yu.H., Yakhvarov D.G., Kargin Yu.M. Coordination Catalysis in Organic Electrosynthesis. Electrochemical Phosphorylation of Organic Halides in the Presence of Samarium Dichloride. Russian Journal of General Chemistry. 1998;68(4) 566-569.

[53] Budnikova Yu.H., Yakhvarov D.G., Kargin Yu.M. Arylation and alkylation of white phosphorus in the presence of electrochemically generated nickel(0) complexes. Mendeleev Communication. 1997;7(2) 67-68.

[54] Budnikova Yu. H., Perichon J., Yakhvarov D.G., Kargin Yu.M., Sinyashin O.G. Highly reactive σ-organonickel complexes in electrocatalytic processes. Journal of Organometallic Chemistry. 2001;630(2) 185-192.

[55] Yakhvarov D.G., Budnikova Yu H., Tazeev D.I., Sinyashin O.G. The influence of the sacrificial anode nature on the mechanism of electrochemical arylation and alkylation of white phosphorus. Russian Chemical Bulletin. 2002;51(11) 2059-2064.